SpringerBriefs in Entrepreneurship and Innovation

Series Editors

David B. Audretsch
Institute of Development Strategies, Indiana University,
Bloomington, Indiana, USA

Albert N. Link
Department of Economics, University of North Carolina Greensboro,
Greensboro, North Carolina, USA

For further volumes:
http://www.springer.com/series/11653

Nicholas S. Vonortas • Phoebe C. Rouge
Anwar Aridi
Editors

Innovation Policy

A Practical Introduction

 Springer

Editors
Nicholas S. Vonortas
Department of Economics
The George Washington University
Washington, D.C., USA

Anwar Aridi
SRI International & The George
Washington University
Washington, D.C., USA

Phoebe C. Rouge
The George Washington University
Washington, D.C., USA

The views expressed in this book are those of the authors and editors and do not necessarily reflect the views of the FTC or any official policy or position of any agency of the US Government.

ISSN 2195-5816   ISSN 2195-5824 (electronic)
SpringerBriefs in Entrepreneurship and Innovation
ISBN 978-1-4939-2232-1   ISBN 978-1-4939-2233-8 (eBook)
DOI 10.1007/978-1-4939-2233-8

Library of Congress Control Number: 2014957311

Springer New York Heidelberg Dordrecht London

Springer is part of Springer Science+Business Media (www.springer.com)

Contents

Contributors

Anwar Aridi SRI International & Trachtenberg School of Public Policy and Public Administration, The George Washington University, Washington, D.C., USA

Benjamin B. Boroughs Center for International Science and Technology Policy, The George Washington University, Washington, D.C., USA

David Feige Center for International Science and Technology Policy, The George Washington University, Washington, D.C., USA

Cherilyn E. Pascoe Center for International Science and Technology Policy, The George Washington University, Washington, D.C., USA

Phoebe C. Rouge Center for International Science and Technology Policy, The George Washington University, Washington, D.C., USA

Nicholas S. Vonortas Center for International Science and Technology Policy & Department of Economics, The George Washington University, Washington, D.C., USA

Daniel Waggoner Center for International Science and Technology Policy, The George Washington University, Washington, D.C., USA

Jeffrey Williams Center for International Science and Technology Policy, The George Washington University, Washington, D.C., USA

Timothy Williams Center for International Science and Technology Policy, The George Washington University, Washington, D.C., USA

List of Figures

List of Tables

—

Chapter 1
Introduction

Nicholas S. Vonortas, Phoebe C. Rouge and Anwar Aridi

This short book provides a quick introduction to important aspects of contemporary innovation policy. It addresses a non-specialist audience interested in quickly building background knowledge, getting familiar with the terminology, and getting an overview of core concerns and debates in this area of policy. The book has its origins in a much more extensive report to the World Bank prepared to impart background information to middle- and upper-level policy decision-makers and analysts as well as stakeholders from industry and universities from developing countries prior to engaging in intensive "how-to" policy training. Our audience also includes upper-level undergraduate and graduate students embarking on the study of innovation policy.

The book is intended as a practical guide to selected issues in innovation policy as they relate primarily to economic growth and development. In preparing the material we have assumed no particular knowledge of the subject matter by the reader and only elementary understanding of economics. The book sets up the policy context and then deals with some of the most important issues in the innovation policy sphere today. It references critical readings on each topic but deliberately avoids bogging down the reader with long reference lists.

N. S. Vonortas (✉)
Center for International Science and Technology Policy & Department of Economics,
The George Washington University, Washington, D.C., USA
e-mail: vonortas@gwu.edu

P. C. Rouge
Center for International Science and Technology Policy, The George Washington University,
Washington, D.C., USA
e-mail: rouge@gwmail.gwu.edu

A. Aridi
SRI International & Trachtenberg School of Public Policy and Public Administration,
The George Washington University, Washington, D.C., USA
e-mail: aridi.anwar@gmail.com

© The Editor(s) 2015
N. S. Vonortas et al. (eds.), *Innovation Policy,* SpringerBriefs in Entrepreneurship
and Innovation, DOI 10.1007/978-1-4939-2233-8_1

1

We do not claim comprehensive coverage of all topics related to innovation policy. Rather than providing a lot of detail, the purpose here is to quickly wrap the reader's mind around basic concepts and quickly enable him/her progress to the topics. For instance, whereas we discuss intellectual property protection and standards, we do not delve into technological paradigms and trajectories and the importance of property rights in these. And, whereas we discuss strategic alliances and high-risk finance, we hardly put the two together to deal with innovative high-risk financing networks. Rather than being comprehensive—an impossible task for a single short book—our aim is to distill and provide adequate information in one place that will prepare a diverse audience to march deeper into more specific topics subsequently.

The more nuanced reader with generalist interest in innovation policy for growth and development will find several other important survey-like books in this field in order to expand beyond the present short book. A partial list would include:

- Jan Fagerberg, David C. Mowery and Richard R. Nelson (eds) (2005) *The Oxford Handbook of Innovation*, Oxford University Press.
- Bronwyn H. Hall and Nathan Rosenberg (eds) (2010) *Handbook of the Economics of Innovation*, Elsevier.
- Chris Freeman and Luc Soete (1997) *The Economics of Industrial Innovation*, 3rd ed., MIT Press.
- Vernon W. Ruttan (2001) *Technology, Growth, and Development*, New York: Oxford University Press.
- Gregory Tassey (2007) *The Technology Imperative*, Edward Elgar.
- Christine Greenhalgh and Mark Rogers (eds) (2010) *Innovation, Intellectual Property and Economic Growth*, Princeton University Press.
- World Bank (2010) Innovation Policy: A Guidebook for Developing Countries, Washington DC: The World Bank.

A set of international organizations also produces streams of very relevant reports including among others the Organization for Economic Cooperation and Development (OECD), the World Bank, the United Nations Conference on Trade and Development (UNCTAD) and Industrial Development Organization (UNIDO), and the World Intellectual Property Organization (WIPO).

This book is comprised of six thematic chapters:

Chapter 2: Fundamentals of Innovation Policy for Growth and Development This chapter provides an overview that initiates the reader quickly into the subject of technology and innovation policy. The chapter begins with a short discussion of the models of economic growth to provide a foundation for understanding how economists view, from a macro-economic perspective, the role that technology and innovation play in the economic growth process. It then proceeds to a more micro-level discussion, beginning with the creation of new technologies (invention) and their commercialization (innovation) and spread (diffusion) across the economy. The chapter then returns to the macro-economic level with a discussion of the relationship between technology and international economic competitiveness.

Chapter 3: University Entrepreneurship This chapter deals with a core sector of the Triple Helix: universities. Specifically, it deals with universities through the prism of entrepreneurship and linkages with industry. The creation of new technologies and new industries rests partially on the transfer of new knowledge to industry, through support of academic research and the movement of scientific talent out to the private sector in the form of trained graduates. The discussion addresses the experience of the United States, the country which is still considered by many as the standard bearer in this respect and the example for other countries to emulate. The chapter focuses on major policy actions and related debates during the past three decades or so in order to flesh out the main points of interest in university-industry relations and the role of the government in trying to foster these by incentivizing higher education institutions to become more entrepreneurial.

Chapter 4: Strategic Alliances/Knowledge-Intensive Partnerships This chapter deals with one of the most important developments during the past few decades: the proliferation of strategic partnerships around the world, especially those based on the production, exchange, and/or use of new technical knowledge. There is little doubt of the centrality of such collaborative agreements across all developed countries and the top tier of developing ones (BRICS+). A strong argument can be made that alliances have a critical role to play in the development and market exploitation of new technologies across all industries and especially knowledge-intensive industries such as those for which information and communication technologies, biotechnology and new materials are important. This chapter deals with this very important issue from the point of view of company strategy and consequent policy implications. It provides a practical guide of the issues involved and illustrates through several cases around the globe.

Chapter 5: Clusters/Science Parks/Knowledge Business Incubators This chapter addresses a major strategic topic in the context of innovation policy: clusters and science parks. These two formations can overlap significantly but are still distinct and thus the chapter is divided into two major parts. Part I deals with the broader concept of clusters (geographical agglomerations of industry to exploit specific locational advantages and spillovers). Part II deals with science parks (geographical agglomerations of industry to exploit proximity with universities and major research institutes). The second part also extends to the incubation of small companies. The chapter is sprinkled with many examples of successful and less successful cases from around the world.

Chapter 6: High Risk Finance This chapter focuses on an absolutely critical aspect of innovation: the transfer of an idea from initial concept to prototype and then to the market. A core component of this process is risk financing, that is, the ability to fund emerging business of higher than average risk. Financial systems around the world struggle with this difficult issue which, nevertheless, has been isolated as of critical importance to development and growth. How does a government deal with the lack of "patient" capital? Venture capital? Investment angels? And so forth. The chapter defines the challenge, provides an overview of the various types of finance for various stages of investment, addresses the important topic of market exit, and then goes into the challenges for emerging markets. The chapter then offers available approaches to supporting high-risk finance by the public sector, offers examples from around the world, and closes with policy recommendations.

Chapter 7: Intellectual Property, Standards This chapter deals with two very important framework conditions of contemporary innovation systems: intellectual property protection and standards. Both these issues—left on the backburner for most of the modern history of industrialization—have been elevated to the forefront due to the arrival of the knowledge-based economy and globalization. Countries that want to be important players in the global economy simply cannot disregard them, even though occasionally they may sound less interesting to some policy decision makers. The chapter summarizes the state-of-the-art in our current understanding of these two topics and relates them to economic development.

Chapter 2
Fundamentals of Innovation Policy for Growth and Development

David Feige

2.1 Introduction

This book deals with technology and innovation and their relationship to economic growth. The emphasis is on policy rather than the underlying economics and the book is designed to be accessible to readers who lack a foundation in economics beyond the principles of the subject. The centrality of economics to an understanding of the underlying processes of economic growth, however, necessitates some discussion of the topic. We have attempted to introduce these concepts in a way that is understandable to the lay reader.

This chapter serves as an overview. It begins with a short discussion of the models of economic growth to provide a foundation for understanding how economists view, from a macro-economic perspective, the role that technology and innovation play in the economic growth process. We will then proceed to a more micro-level discussion, beginning with the creation of new technologies (invention), and their commercialization (innovation) and spread (diffusion) across the economy. We will then return to the macro-economic level with a discussion of the relationship between technology and international economic competitiveness.

It is worthwhile first to define some basic terms so that the reader understands the vocabulary used throughout the book. The words "science and technology" are frequently used together but their separate meanings are sometimes lost in the process. Similarly, the terms "technology" and "innovation" are sometimes used interchangeably. For our purposes, *science* is the systematic search for new knowledge. *Technology* is the application of that knowledge to the production process. *Innovation* can be distinguished from technology by understanding that technology is only one way to innovate. Although it is the most common form of innovation in developed countries, there are other forms of innovation including innovations in

D. Feige (✉)
Center for International Science and Technology Policy, The George Washington University, Washington, D.C., USA
e-mail: dfeige@gwu.edu

© The Editor(s) 2015
N. S. Vonortas et al. (eds.), *Innovation Policy,* SpringerBriefs in Entrepreneurship and Innovation, DOI 10.1007/978-1-4939-2233-8_2

marketing or organizational form. Other terms will be introduced in the course of this chapter as well.

This book has a strong policy focus. As such, the assumption that underpins its content is that policymakers can intervene (productively) to encourage the production and use of new technologies. While the existence of market failures suggests a useful role for governments, it is true that not all government intervention is helpful and can occasionally be counterproductive. We will attempt throughout to highlight what we believe to be the appropriate role of government in encouraging and accelerating the process of technology creation, commercialization, and diffusion.

How policy affects innovation

2.2 Models of Economic Growth

This section provides an overview of some of the primary economic growth theories and the way they have evolved over time to account for the role of technology and innovation in the economic growth process. It provides context for more policy-oriented sections to follow. We define economic growth as a sustainable increase in GDP per capita. The section will explore neoclassical growth theory; endogenous growth models; and evolutionary models; followed by a brief discussion of the convergence hypothesis.

2.2.1 The Neoclassical Growth Model[1]

The neoclassical growth model, also known as the "Solow-Swan" model, was probably the first modern model of economic growth to explicitly recognize the role of technology as a central driver of economic growth. It is associated most closely with Robert Solow, who observed in 1957 that a large part of U.S. economic growth was unexplained by the contributions of capital and labor, the two factors that characterized earlier models. Solow (1957) attributed this unexplained element to technological change and referred to it as Total Factor Productivity, or TFP (Moses Abramowitz referred to it as "the measure of our ignorance" in recognition of the fact that we have very little understanding of the myriad factors that contribute to it and the degree to which each does so). In Solow's model, only growth in technology can result in sustainable economic growth. Importantly, Solow's model assumes that technology is produced exogenously (outside of the model). We shall see in a moment that this has been a key point of contention with some of the more recent models.

Additionally, the model identified a "steady-state" rate of growth, or the growth rate that a country could theoretically sustain in the long term. "Over-performing" countries, or those above the steady-state rate of growth, would inevitably regress

---

[1] Sect. 3.1 and 3.2 draw on Greenhalgh and Rogers (2010).

to that rate of growth; while those countries performing at a sub-optimal level (a level below their steady state) would naturally increase their growth rate until they reached that sustainable rate. An important implication of Solow's model, then, is that it suggests that underperforming countries will grow faster than better performing economies do. That is, the poorer a country is (in terms of GDP per capita) the more quickly it would grow relative to wealthier ones. This suggested the inevitability of "convergence", or the gradual catch-up of poorer countries to richer ones.

2.2.2 Endogenous Growth Theories

Solow's model began to receive serious challenges in the 1970s as some of its key assumptions appeared to conflict with observed reality. The first was its assumption that technology was produced outside of the model, which seemed inconsistent with the fact that much invention and innovation is part and parcel of the economic system and is very much determined by the everyday decisions of the economic units in this system. Second, the model continued to under-explain actual observed rates of economic growth. And third, while some countries appeared to be converging others appeared to be diverging from the leading economies. An important set of these challenges coalesced into what is known as the endogenous growth theory (or New Growth Theory), most closely associated with Paul Romer (1986).

Endogenous growth models made three key assumptions distinct from Solow's model. First, they assumed that the production of technology is endogenous (internal), rather than exogenous (external), to the model. That is, they recognized the explicit role of economic units such as firms in the production of new technologies. Second, they assumed that knowledge could "accumulate"; that knowledge is a cumulative process that could be maintained and added to over time. Finally, they also assumed that knowledge "spills over"; that knowledge produced by one firm may be useful to others. Further, this process is inter-temporal; that is, firms can benefit from knowledge that was produced by other firms at an earlier point in time.

The endogenous growth model has important implications. While Solow's model assumed that capital has diminishing returns (that is, each additional dollar of capital results in a lower amount of additional output, everything else constant), in Romer's model, although individual firms may face diminishing returns to capital, the economy as a whole does not. This suggests that growth is possible in the long run and contrasts with Solow's prediction that growth could not be sustained at levels above their "steady state". While other variations on the endogenous growth model exist (see, for example, Lucas 1988), Romer's remains the most widely known.

2.2.3 Evolutionary Economics

Many of the ideas embodied in the endogenous growth models had already been discussed previously in a loose coalition of economic thought called evolutionary economics, such as the ideas on the nature of knowledge, the way it accumulates, and the possibilities for systemic learning and for increasing returns. However, evolutionary economics also challenged some of the basic concepts of neoclassicism which also continued in endogenous growth and is thus considered a separate (and challenging) school of thought.

Evolutionary economics is inspired by biological processes and focuses principally on two ideas (Verspagen 2005). The first is that firms are "chosen" by the market based on their ability to adapt to changing circumstances. The second is that innovation simultaneously (and continuously) introduces novelty into the system, effectively creating a "moving target" that firms need to adjust to. A third can be added regarding the way firms make decisions: rather than maximizing profits (which requires a huge amount of information), they develop and follow "sticky" routines and maximize "satisfying" behavior (i.e., make their owners feel happy with their investment). The constant interaction between the ever-changing system and the firms that inhabit it determines the "winners" that emerge. Importantly, these outcomes are difficult to predict. One strain of evolutionary economics postulates that technological development (and therefore economic growth) is dictated largely by technological trajectories or paradigms, which determine the parameters within which technology will advance for extended periods of time. These provide the context for specific innovations which "cluster" in time because a series of incremental innovations closely follow a radical one. The largest and most significant of these innovations may be so-called General Purpose Technologies, or GPTs, that are characterized by their broad application throughout the economy, such as ICT, biotechnology, or new materials.

There are two key distinctions between evolutionary economics and endogenous growth theory. First, endogenous growth theory assumes that firms are aware of the entire range of potential technologies and as such can "jump" from one technology to another as technologies prove themselves to provide a more profitable set of outcomes. Evolutionary economics, on the other hand, suggests that firms tend only to be aware of technologies very close to their current technology and are thus not necessarily able to take advantage of new technologies as they present themselves. Second, endogenous growth theory assumes "weak uncertainty" associated with policy choices (that is, the range of outcomes related to a policy choice are known but the specific outcome that will result is not); while evolutionary economics adheres to "strong uncertainty" (that policymakers are not even aware of the full range of outcomes). Therefore, while endogenous growth theory assumes that a series of policy levers can be pulled to result in a fairly predictable outcome, evolutionary theory suggests it is much more difficult to know what the outcome of specific policies will be.

2.2.4 The Convergence Hypothesis

We close this section with a brief word on the convergence hypothesis. It was mentioned earlier that Solow's model predicts convergence; but that we observe a combination of convergence and divergence. That is, some countries appear to be converging with (catching up to) the leading economies, while others appear to be diverging from them. A concise characterization of the convergence hypothesis was given by Baumol et al. (1989). When the productivity level of one or more countries is substantially superior to that of a number of other economies, largely as a result of differences in productive techniques, then laggard countries that are not too far behind the leaders will be in a position to embark upon a catch-up process. Many of them will actually do so. The catch-up process will continue as long as the economies approaching the leader's performance have a lot to learn from the leader. As the distance among the two groups narrows, the stock of unabsorbed knowledge will diminish and even approach exhaustion. The catch-up process will then weaken or even terminate unless some other unrelated influence comes into play. Meanwhile, those countries that are so far behind the leaders that find it impractical to profit substantially from the leaders' knowledge will generally not be able to participate in the convergence process at all. Many such economies will find themselves falling further behind, widening the gap between wealthy and poor nations.

The convergence hypothesis was empirically tested and debated over the years. According to Baumol et al. (1989), a country's ability to "converge" with leading economies is a function of (1) capital accumulation, (2) technological innovation, and (3) imitative entrepreneurship (which borrows ideas from abroad and adapts them to local circumstances).

Abramowitz (1986), on the other hand, highlights the role of social capabilities (effective institutions, including incentives and markets) in determining which countries are best able to close the gap (converge) with countries at the technological frontier. He adds to social capability the importance of "technological congruence", that is, the transferability of the leader's technology to follower countries. Essentially, countries that have developed sufficient capabilities and technological congruence are able to close the gap with the leaders due to the fact that they are able to copy and absorb the technologies the leaders have produced. As the stock of unabsorbed knowledge and technology shrinks, the pace at which convergence happens slows until it eventually comes to a halt as there are no more technologies to copy. (At that point, countries that have caught up can continue increasing their growth rate above that of other technological leaders only by producing their own new technologies). Those countries, however, that lack the capabilities to "understand" and therefore copy and absorb the technologies produced by the leaders, will fall further behind, resulting in divergence from the leaders.

Importantly, the convergence hypothesis predicts a different set of outcomes from those produced by Solow's model. While Solow assumes that convergence is inevitable, convergence theory suggests that it is not; and that good policy can play an important role in determining whether a country takes the path of convergence or of divergence.

2.3   Technology Creation (Invention)

We have now provided some context for the importance of technology in the economic growth process. We proceed in the next three sections to a discussion of how the growth of technology is nurtured. This section focuses on the creation of new technologies. We look first at the mechanics of technology creation. This is followed by a discussion of the rationale for government intervention in the support of research; and concludes with two sections that look more closely at issues of specific interest to policymakers.

2.3.1   The Research Chain

The process of technology creation is often divided into three stages: basic research, applied research, and development (although in reality the lines between the three are blurred). *Basic research* is distinct from applied research in that it is conducted without consideration for a specific application. *Applied research*, on the other hand, is undertaken with a specific need in mind. *Development* is the design, construction, and testing of prototypes of new products and processes. Research is critical because it is the foundation for technology (which, it will be recalled, was defined in Sect. 2.1 as the application of new knowledge to the production process). Technology, in turn, is central to productivity growth, as discussed in Sect. 2.3.

2.3.2   Economic Arguments for Policy Intervention in Research Activity

Most arguments for public intervention in research relate to the more basic and generic aspects of research; as the government is generally considered to be too far removed from the market to play a useful role in applied research.

There are two primary economic arguments that justify public intervention in research activity. The first rests primarily on the theory of *market failures*. This argument suggests that:

- The social returns related to research activity outweigh private benefits, implying that private sector actors are likely to under-invest in research; and
- A high level of uncertainty characterizes R&D and innovative activity, which can be only partly insured.

In addition, market failures can arise due to the fact that certain investments can be made only at significant scale; and as a result of information asymmetries between the parties conducting research and those funding it.

The second economic argument is based on *system failures*. One case of this is when introduction of an initial technology leads to "lock-in" along a sub-optimal technological trajectory—such as, arguably, fossil fuels today. A second case, discussed in greater depth in Sect. 2.6.4.2, relates to the need for coordination among institutional actors in order to promote the diffusion of innovations. A third case in which the government can play a useful role is in making strategic R&D investments both within technology cycles and in managing the transition from one technology life cycle to another. In addition, public intervention can also be important in developing human capital for the purpose of promoting absorption of technology.

2.3.3   Issues of Interest to Policymakers

2.3.3.1   Intellectual Property Rights (IPRs)

One of the most widely discussed policy issues with respect to the creation of new technologies is that of intellectual property rights, or IPRs. IPRs encompass patents, trademarks, copyrights, and trade secrets; these are discussed more extensively in Chap. 7 of this book. We will focus briefly here on patents. Patents in effect grant the inventor a temporary monopoly, thereby allowing them to capture all of the economic benefits from their invention over a limited period of time; in exchange for the inventor's agreement to put all knowledge related to the invention into the public domain. The patent system is therefore an attempt to solve the appropriability problem addressed above.

Several concerns have been raised, however, with respect to the patent system. One relates to the duration of patents and whether it should be uniform across sectors and technologies given the great differences among them. A second involves questions about whether the exercise of some of the rights associated with owning a patent may in fact discourage, rather than encourage, invention. One example is the practice of obtaining patents (with no intention of using them) for the knowledge surrounding an invention a firm currently holds a patent to, thereby preventing other firms from "inventing around" the patent that the firm hopes to exploit. A third issue concerns the cost of the patent system and whether that disproportionately benefits larger firms relative to smaller ones. A fourth involves the length of time necessary to obtain a patent, which may make the technology to be covered by the patent obsolete by the time patent approval is granted. Finally, lax IPR systems in many developing countries have also raised criticisms from more developed countries. In many cases these have been established specifically to promote the diffusion of technologies (discussed in Sect. 2.5) in countries that lack the capacity to produce leading-edge research; but this remains an ongoing subject of controversy.

It is also unclear to what extent patents are central to the decisions of firms to produce (applied) research. Research shows that firms outside of the pharmaceuticals and chemicals sectors rely on patent protection to only a very limited extent (or not at all) to protect their inventions,[2] preferring instead to establish first-mover advantage or the development of complementary capabilities to create a market

---

[2] Mansfield's work (referenced in Cohen 2010, pp. 182–183)

position that cannot easily be imitated. Firms also in some cases choose not to patent in order to avoid having to put knowledge into the public domain (preferring to resort to trade secrets instead).

2.3.3.2 R&D Composition

Another (often overlooked) issue of interest to policymakers is the composition of R&D spending. Many countries have attempted to target an "optimal" level of R&D spending (3 % of GDP, which was chosen by the European Union in their 2020 growth strategy,[3] seems to be a particularly common target for developed economies, although Korea and a few others have higher stated targets), but have neglected any attention to the split between basic and applied research spending. As noted earlier in this section, while applied research is the basis for products and services that can be commercialized in the near future, basic research plays a critical role in producing the foundation for the technologies that will drive competitiveness in the future. The amount of funding devoted to applied research (most of which is funded by companies) relative to basic research (most of which is funded by governments) typically increases as countries develop. However, there are frequently voiced concerns that insufficient resources are being devoted to basic research activities, thereby potentially compromising a country's future competitiveness. Of additional import is the destination of R&D funding; whether it is oriented toward defense application, for example, or designated for uses that are more likely ultimately to have commercial application.

2.3.3.3 Non-Linear Research Models

We have mentioned that the neat division of research activity into basic research, applied research, and development is an oversimplification of the way that new technologies are developed. This is typically referred to as the *linear model*, and implies that the process of technology creation occurs in a predictable order. In reality, the process is often more iterative than linear. The publication of *Pasteur's Quadrant*, by (Stokes 1997) epitomizes this thinking; calling into question the linear model (basic research leads to applied research which in turn leads to development, production and marketing of new products) while suggesting that the process involves a stronger feedback mechanism (from the market to research) than the linear model envisioned and could be initiated at multiple points in the "research chain". This fact has important policy implications as it suggests that governments will need to strike a balance between "supply-led" policies (in which R&D funding is typically driven by the missions of public organizations) that characterize the linear model and "demand-led", or user-driven, policies, such as those promoting market innovations, that recognize that the end markets play an important role in informing the research that is conducted.

---

[3] As cited in Albu (2011).

2.3.4   Policy Tools Available to Support Basic Research

Governments can tweak the intellectual property system to obtain desired outcomes; for example, the Bayh-Dole Act in the U.S., which granted the rights to intellectual property produced by universities with federal funding to the universities themselves, has probably incentivized universities to produce more research of value than they might have in its absence (more on this in Chap. 3). However, governments have other tools at their disposal as well. We will mention two; direct support to R&D and tax incentive programs.

Direct support (generally in the form of grants and contracts) ranges from about 20 % of total research expenditures in East Asian countries such as Korea and Japan to up to 50 % in select European Union countries (the U.S.'s federal share is about 33 % of total research expenditures) to higher shares in countries like Brazil (Steen 2012). Much of the public funding in developed countries tends to be directed to universities, which, for example, conduct over half of all basic research in the U.S. Such direct funding for research offers policymakers the advantage of being able to choose where the funding goes while still keeping at some distance from the market.

An alternative to direct support is indirect support through the provision of tax incentives to companies. Such incentives provide matching funds to companies for every dollar of research that they conduct; or for every dollar of research they conduct above a certain baseline (usually determined by past R&D investments by the company). Tax incentives are controversial because of the difficulties associated with linking them to actual increases in company R&D spending. Most research suggests that there is approximately a 1:1 ratio between government spending and research funding allocated; that is, companies increase their total R&D spending by, on average, exactly the amount they receive from the government; which may seem an inefficient subsidy mechanism in catalyzing additional R&D investment.

An additional policy option available to governments is the support of collaborative research partnerships. These partnerships may take the form of public-private arrangements (such as those between governments and private companies) or private-private arrangements (which encourage companies to work together, often through strategic alliances or joint ventures, to produce basic research). This is the subject of Chap. 4 of this book.

2.4   Commercialization of New Technologies (Innovation)

We now turn to a discussion of the commercialization of new technologies, typically the idea associated with innovation. Only a small percentage of all inventions actually become innovations; that is, very few inventions actually find commercial application. Most research suggests that only about 2 % of all patents find commercial use. As not all inventions are patented, this is only a representative figure; but does provide some sense of the limited number of new technologies that are created that actually make it to market. Because of this, it is important to understand the dynamics of the commercialization process.

2.4.1 Commercialization and Large Firms

Schumpeter, J. (1942) and his followers at one time asserted that large firms are more capable of generating innovations than small firms are. While extensive research since then has shown this to be inconsistent with the evidence, large firms do play a very important role in commercializing technologies in certain industries, including for instance highly capital-intensive industries such as pharmaceuticals and chemicals and industries requiring the integration of complex products such as automobiles, aircraft, and military equipment. Possessing access to many resources, large firms account for the majority of absolute spending on R&D in the US. In addition, large firms are also the source of numerous spin-offs (discussed in Sect. 2.4.2), thus playing a central role in the innovation ecosystem.

2.4.2 Commercialization and Entrepreneurship/Small Firms

Entrepreneurship was initially largely ignored in discussions of national systems of innovation (discussed in Sect. 2.6.4.2) but has, in the last decade, become a priority in policy circles. Of most interest for this book is the category of entrepreneurs we refer to as *growth entrepreneurs* (also referred to as "opportunity entrepreneurs"), which we define as individuals or teams of people who exploit a previously unidentified or unexploited business opportunity. We distinguish this group from *necessity entrepreneurs*, most commonly found in developing countries, who have turned to entrepreneurship as a livelihood only in the absence of other job opportunities. Within the category of companies set up by growth entrepreneurs, the most important sub-set is R&D-intensive companies. In developed countries this group contributes disproportionately to job creation and innovation and is therefore of great interest to policymakers. Only between 2–4 % of all small and medium sized enterprises (SMEs) can be classified in this group at any point in time. The entire "Research Stairway", and the percentage of firms that fall into each category of research intensity, is illustrated in Fig. 2.1.

Another, largely overlapping, sub-set of companies set up by growth entrepreneurs is the so-called "gazelles", those enterprises that have demonstrated sustained, above average growth in profits. According to a recent report, only about 4 % of respondents fell into this category; but accounted for about 40 % of new job creation in the United States (Endeavor 2011).

While entrepreneurial activity has frequently been attributed to the somewhat mystical qualities of a few gifted or creative individuals, the reality is that it is driven by the interaction of these individuals with the system within which they operate. Thus, the concept of "National Systems of Entrepreneurship" (Acs et al. 2013) has arisen in recognition of this systemic element to the "creation" of entrepreneurs. This recognizes that policymakers have a role in creating an environment supportive of those individuals who have entrepreneurial aspirations, a subject that will be discussed in greater depth in Sect. 2.4.3.

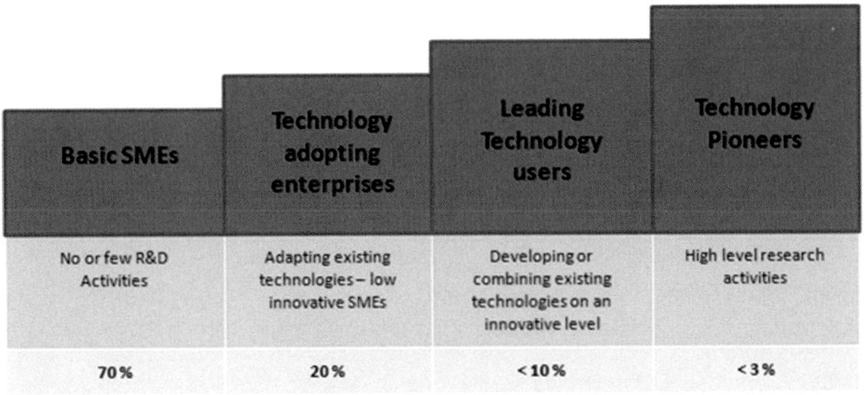

Basic SMEs	Technology adopting enterprises	Leading Technology users	Technology Pioneers
No or few R&D Activities	Adapting existing technologies – low innovative SMEs	Developing or combining existing technologies on an innovative level	High level research activities
70%	20%	<10%	<3%

Fig. 2.1 The research stairway (EURAB 2004)

Entrepreneurs can arise, of course, in any industry. Within the context of our discussion of technology and innovation, we are particularly interested in the role that entrepreneurs (and small firms) play in commercializing new technologies. In line with Schumpeter's hypothesis with respect to innovation and firm size, it was at one time believed that large firms were more innovative than small ones. However, more recent research suggests that, although large firms have an advantage innovating in certain industries (as mentioned in Sect. 2.4.1) small firms are, on average, disproportionately responsible for innovation as a whole (Acs and Audretsch 2001). Their relative advantage seems greater when it comes to radical innovation.

They do so in primarily two ways. One is by commercializing research performed in universities; this may happen either when an inventor decides to commercialize his/her own research or through a licensing arrangement. The second is through "spin-offs" from existing firms; a common phenomenon is that an entrepreneurial individual produces an invention within the context of a larger firm to which they assign more value than the firm itself does. In such cases, the entrepreneur may leave the firm, taking their invention with them, and commercialize it under the auspices of a new company (Auerswald and Branscomb 2003). Such practice has been, in fact, institutionalized in certain large companies which sense a window of opportunity on the one hand—spin-off firms that may succeed may be folded back into the corporation later on—while dissipating internal conflicts on the other. In this way entrepreneurs play a key role as conduits of knowledge spillovers, addressed in our discussion of endogenous growth theories in Sect. 2.2.2. While several large companies are attempting to set up innovative units internally to stem the flood of talent leaving the firm and to capture more of the value of such innovations as they come online, such efforts have met with mixed success.[4]

---

[4] The early example of Xerox's PARC and the current Skunk Works of Lockheed Martin are cases in point.

2.4.3   Policy Interventions Supporting Entrepreneurship and Small Businesses

The focus of policy with respect to commercialization has focused primarily on support to entrepreneurship and small businesses in recognition of their central role in the innovation process. We will touch on a few support mechanisms here; including (1) financing and technical assistance programs (often provided through science parks and business incubators), (2) government procurement, and (3) National Systems of Entrepreneurship.

2.4.3.1   Finance and Technical Assistance

Financing for small enterprises has long been of interest to policymakers. Particular attention has been paid to the so-called "valley of death" that frequently engulfs small enterprises between basic and early applied research, on the one hand, and initial innovation and commercialization, on the other. This refers to the funding gap that exists that is not addressed by either the typical public sector programs supporting research, by angel investors, or by venture capital; thus resulting in the vast majority of small business failures. This subject is more thoroughly covered in Chap. 6 of this book, but the Small Business Innovation research (SBIR) program in the United States is a well publicized attempt by the public sector to address it. The SBIR is generally regarded as a fairly successful model for financing early stage innovation and has been adopted by several countries around the world. Technical assistance programs are another form of non-financial support and may include basic business skills training, help with marketing or product development, or linkages to domestic or export markets. These services are often provided in the context of a business incubator, which provides access to both financing and technical assistance in addition to physical space for the enterprise to operate.

2.4.3.2   Government Procurement

Government procurement is another, probably underutilized, tool that governments have at their disposal to encourage innovative activity among small firms. The military has often played an important role in sourcing leading-edge technologies that ultimately found commercial application, especially in developed countries (semiconductors is a widely cited example); and much of this work was contracted through small businesses. Small firms can similarly play a role in other, non-defense industries through set-aside grants designed to source innovative products or to source technologies specifically from small firms.

2.4.3.3 National Systems of Entrepreneurship

Brief mention was made in Sect. 2.4.2 of the concept of National Systems of Entrepreneurship (Acs et al. 2013). This approach recognizes that the creation of systematic innovative entrepreneurship requires a holistic "ecosystem" approach (including funding, mentorship opportunities, market linkages, and a catalytic environment with respect to business rules and regulations); and that governments have an important role in filling the "gaps" that exist in the entrepreneurial ecosystem. This suggests that regulations should be harmonized and oriented toward the support of entrepreneurial activity; and that this should be accompanied by the development of technological infrastructure as well as efforts to change national cultures that often hinder entrepreneurial activity. These may include media campaigns to try to change the attitudes of individuals toward entrepreneurship as a career as well as changing the attitudes of society toward entrepreneurs.

2.5 Technology Diffusion

Technology diffusion refers to the spread of technology throughout an economy. Diffusion is the principal determinant of the contribution that a specific technology makes to economic growth. In this section we outline the diffusion process before proceeding to specific policy interventions that can be used to affect the speed with which diffusion takes place; and conclude with a discussion of international diffusion processes and the role that multinational companies (MNCs) play in that process.

2.5.1 The Diffusion Process

One of the key observations made with respect to the diffusion process is that diffusion does not happen suddenly, but is rather a gradual process that begins haltingly, speeds up once it hits a "takeoff point", and then slows down as the market for the particular innovation saturates. (This pattern takes the form of an "S" shape). Additionally, new technologies are typically adopted by different people at different times. Widespread diffusion of a given technology depends largely upon the extent to which creators of a new technology are successfully able to reach innovators and early adopters, who provide important feedback on the technology before it reaches a wider audience. The S-shaped diffusion curve, overlaid on the adoption curve, is illustrated in Fig. 2.2 below.

Importantly, however, different innovations diffuse at different rates. The speed of diffusion is determined by many factors, among which include (1) the degree to which the new technology represents an improvement over the old, (2) the cost of

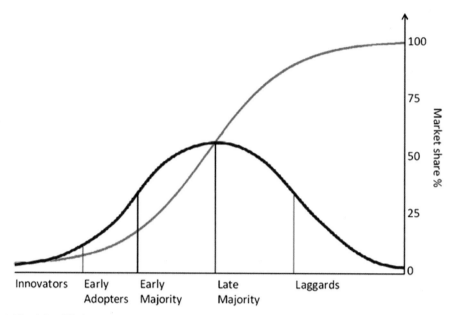

Fig. 2.2 Diffusion and adoption curves. ("Everett Rogers", Wikipedia: The Free Encyclopedia. 2014)

adopting the innovation (including both the technology and any complementary technologies that must be purchased to accommodate it), (3) improvements made to the technology in the course of the diffusion process, (4) the degree of uncertainty about the new technology, (5) the existence of complementary technologies, (6) the existence of "network effects", which arise when each user benefits from the addition of new users (think, for example, of cell phone networks), and (7) the extent to which improvements are made to the incumbent technology in an effort to stave off its replacement (Hall 2005; Rosegger 1986; Stoneman and Battisti 2010).

2.5.2 Diffusion Policy

Governments can play an important role in either speeding or slowing diffusion. One excellent example is the patent system; stronger patent protection (typical in more developed countries) often slows the diffusion process by the granting of a temporary monopoly over new technologies. Looser patent protection in developing countries, on the other hand, can play an important role in speeding diffusion by allowing for easy imitation of existing technologies. It is important to recognize the trade-off here; stronger patent protection creates an incentive system more aligned with the creation of new technologies, while looser protection spurs diffusion. Thus, choice of a patent regime depends greatly on a country's position as a technology leader or a technology laggard, with leaders adopting stronger patent protection and laggards favoring weaker protection. These different perspectives have given rise to a heated debate internationally about the extent to which developing countries should be able to establish weak IPR regimes.

Governments can also promote diffusion of certain technologies by setting standards; or by establishing regulations that encourage the uptake of certain technologies. A common example is the importance of emissions standards in promoting the diffusion of alternatives to carbon. On the other hand, governments can also slow the diffusion process, for example through the slowness of regulatory change. Finally, governments can promote diffusion of certain technologies over others by "picking winners" (favoring one technological solution over another) through the use of subsidies such as tax breaks. This is generally considered a more inefficient mechanism for promoting diffusion than the use of regulations that do not favor specific technologies. Regulation tends to be more effective at allowing the market to determine the technologies that ultimately win out.

2.5.3 International Diffusion

One additional area of interest is diffusion across borders. Multinational corporations (MNCs) are the primary agents of international diffusion, primarily through foreign direct investment (FDI) but also through other mechanisms, such as exports. While FDI is an important source of both job creation and exports in receiving (host) countries, the principal attraction of foreign investment is the access that it provides to the "black box": technologies that can be exploited by the host country for its benefit. FDI can result in technology transfer, which refers to the intentional sharing of technologies with local firms by MNCs; or technology spillovers, which are unintentional or incidental to the investment.

However, the extent to which these occur is highly dependent on the absorptive capacity of host countries—including the technological gap between the MNC and host country firms, and the extent to which the country and its firms conduct research and development to enable them to better "understand" technologies produced outside of the country—as well as firm strategies regarding the nature of the FDI they are making. FDI can be divided into (1) asset-augmenting FDI, which seeks to exploit and leverage local capabilities (such as the acquisition of either R&D facilities or technological capabilities) and is more likely to generate host-country benefits; and (2) asset-seeking FDI (such as investment in extractive industries), which aims principally to take advantage of local natural resources or to gain access to local markets, and is basically exploitative in nature.[5] The nature of that investment, though, tends to be determined by the extent to which local (host country) absorptive capacity[6] exists (as well as host country policy); that is,

---

[5] Dunning (1993) originally defined four categories of FDI (market-seeking, resource-seeking, efficiency-seeking, and strategic asset-seeking); these have been condensed in the more recent literature to these two.

[6] Absorptive capacity was originally defined by Cohen and Levinthal (1989, p. 569) as "a firm's ability to identify, assimilate, and exploit knowledge from the environment".

absorptive capacity and firm decisions regarding the nature of their investment are co-determinant.

Governments thus have an important role to play in building local capacities such that a country is better positioned both to attract the investment it seeks and to access the technologies that will allow it to climb up the value chain. It is worth noting here the temptation that policymakers face to establish local content requirements. Although these would seem on the surface an effective way of assuring the transfer of technology, they often tend to drive investment away, thus precluding local firms from benefiting from the presence of foreign firms. The exception to this is larger countries (such as India, Brazil, or China) which boast attractive enough local markets allowing them a stronger negotiating position *vis-à-vis* MNCs.

One additional point should be made here. The diffusion process in developing countries often hinges on entrepreneurial individuals who adapt and adopt technologies from abroad. This process, in which individual entrepreneurs are the agents through which a country identifies the sectors and sub-sectors in which it holds a current or potential competitive advantage, has been dubbed "self-discovery" by Rodrik and Hausmann (2002) and "imitative entrepreneurship" by Baumol (1968). Because of the simultaneous nature of innovation and diffusion within this process, it becomes increasingly difficult to distinguish one from the other. While innovation and diffusion are concomitant processes to a certain extent in developed countries as well, the lines become further blurred in developing countries because most inventions do not originate there.

2.6   Technology, Innovation, and International Economic Competitiveness

Technology and innovation are central to economic competitiveness. In more developed countries, competitiveness is primarily a function of their ability to develop new technologies; while in developing countries competitiveness is dependent upon their ability to utilize existing technologies. In this section we first review the definition and ways of assessing competitiveness. We then proceed to a discussion of the decentralized nature of the drivers of competitiveness; before concluding with a discussion of a pair of particularly topical issues with respect to policy around innovation and competitiveness: clusters and systems of innovation.

2.6.1   Defining Competitiveness

Competitiveness is a frequently used (and almost as frequently misused) term in policy circles. While the term carries different meanings at the firm, industry, and national levels, for the purposes of this report, we will define competitiveness as the ability of a country to generate sustained increases in productivity resulting in a high and rising standard of living for its people. Productivity, on the other hand, is a measure of the efficiency with which a country is able to convert its inputs (resources) into outputs (products and services) that have value on domestic and/or international markets.

Michael Porter (1990) draws a distinction between what he refers to as "comparative advantage" and what he calls "competitive advantage". Comparative advantage arises from the exploitation of "inherited" or "basic" factors of production, such as fertile land or cheap labor. Competitive advantages, on the other hand, are "created"; they include advanced infrastructure and skilled human resources, and require some effort (investment) on the part of the country to produce. Only the latter can generate sustained increases in a country's standard of living because only enhancements to a country's competitive advantage allow for increases in productivity, which translate into increases in wages. Technology and innovation are central to a country's capacity to generate competitive advantage.

2.6.2 Assessing Competitiveness: The Global Competitiveness Index

The Global Competitiveness Index (GCI) is a composite index published annually by the World Economic Forum. Contrary to popular misconceptions, the GCI does not actually measure "competitiveness". (Competitiveness, as noted above, is defined as productivity, and thus can be measured already without the help of the GCI). Rather, it is an attempt to *explain* differing levels of productivity among countries and to predict the ability of a country to achieve high and rising levels of productivity (and, by extension, national prosperity) in the future. It therefore uses a model (hypothesis) about the factors (indicators) that contribute to a country's level of competitiveness and their relative importance; and then uses a combination of hard data and a survey of company executives to assign scores to a country on each factor. The factors are categorized into a set of broad "pillars" that include institutions; infrastructure; macroeconomic stability; health and primary education; higher education and training; goods market efficiency; labor market efficiency; financial market sophistication; technological readiness; market size; business sophistication; and innovation. The twelve pillars are further condensed into three categories that are roughly aligned with the stages of development of a country (defined by GDP per capita). By identifying the areas in which a country has demonstrable weaknesses, policymakers are able to make targeted improvements to address those shortcomings. Although it suffers from a number of imperfections, both in the content of the related survey as well as its sampling techniques, the GCI does show a broadly positive correlation with GDP per capita levels and remains the most popular tool for assessing and explaining a country's overall competitiveness.

For our purposes, it is probably most important to note that the GCI distinguishes among three types of economies, each of which demonstrate a set of characteristics that also (with a few exceptions) correspond roughly to GDP per capita levels. *Factor-driven economies* tend to be based heavily on the exploitation of basic factors of production and are characterized by primary product exports. *Efficiency-driven*

Global Competitiveness Index		
Basic requirements subindex	**Efficiency enchancers subindex**	**Innovation and sophistication factors subindex**
1. Institutions 2. Infrastructure 3. Macroeconomic environment 4. Health and primary education	5. Higher education and training 6. Goods market efficiency 7. Labor market efficiency 8. Financial market development 9. Technological Readiness 10. Market size	11. Business sophistication 12. Innovation
Key for **factor-driven** economies	Key for **efficiency-driven** economies	Key for **innovation-driven** economies

Fig. 2.3 Pillars of the global competitiveness index. (Adapted from Schwab 2013)

economies are characterized by their ability to produce existing products efficiently (generally by deploying existing process technologies) but tend not to produce new products or processes. *Innovation-driven economies*, on the other hand, are the source of significant new products and processes. Section 2.5.3 on the international diffusion of technology discussed the way that MNCs make decisions regarding where they locate specific activities within global value chains. We can think of these categories (while they obviously require a certain degree of simplification) as the rough contours that determine where those activities are located. They are therefore both reflective of, and a determinant of, a country's ability to generate higher value addition.

The 12 key "pillars" of the Index, as well as the stage of development with which each is most closely associated, is summarized above (Fig. 2.3).

It is worthwhile noting that the depiction above assumes that progression through the three stages (factor-driven through innovation-driven) is essentially linear in nature; and that a country must develop its basic requirements prior to developing its efficiency enhancers; which must precede development of its innovation and sophistication factors.

2.6.3 Competitiveness as a Decentralized Process

The fact that most measures of competitiveness use the national level as the economic unit of measurement obscures that fact that competitiveness is in large part a subnational process, dependent on the specific factors of production located in regions within countries. This is particularly true in larger countries. The mechanics of this process are a matter of some dispute, however.

There are two principal schools of thought regarding the way that decentralized economic growth happens. One of these, popularized by Michael Porter as "industrial clusters" but is an idea that dates back to Alfred Marshall's work on agglomeration economies from the late 19th century, is discussed in greater depth in Chap. 5 of this book. For now, it suffices to note that cluster-based approaches assume that competitiveness is generated by inter-firm and inter-industry effects; that is, the dynamic interaction of firms in a given sector and the related industries and institutions that support them. Many of the most dynamic clusters are based on science and technology. These include not only Silicon Valley but the ICT cluster in Bangalore, India, life sciences in Boston, Massachusetts, and aerospace in Sao Jose dos Campos, Brazil.

A competing theory arises from the field of urban economics, first postulated by Jane Jacobs (1969). This school of thought asserts that the agglomeration of people, not industries, drives productivity increases; and that knowledge spillovers occur not just within industries but across them. This field focuses on urban areas as hotbeds of innovation and stresses the importance of diversification rather than specialization.

The fact that competitiveness is largely determined at the sub-national level suggests that policymakers at the local and regional level have a critical role to play in this process; and that it is necessary that policymakers at the national level coordinate their efforts closely with local and regional officials.

2.6.4 Policy Implications

2.6.4.1 Clusters and Technology Parks

Many regions globally have made cluster policy a staple of their economic development strategies. Similarly, technology parks, which have been defined as "mixed-use real-estate developments built close to universities that seek to encourage industry-university knowledge transfer" have become a popular tool for driving regional development. As we have devoted an entire chapter to clusters and technology parks in this book, we will not go into great detail here. There are two issues worth highlighting, however.

First, many governments have become wedded to the idea of generating "high-tech" industries in their area, and have gone to great lengths to try to "create" such industries. Most regions that have attempted to create industries from scratch have failed in doing so sustainably. It is generally better policy to build on a region's existing advantages than to attempt to build something with no prior history in the area. Second, while it is true that technology and innovation are critical drivers of productivity growth, it is the view of the authors that disproportionate focus has been placed on developing "high-technology" industries. Such a focus ignores the fact that any industry, irrespective of how it is categorized, can demonstrate high levels of productivity and therefore drive economic growth. Technology can be viewed as an "enabler": firms in "low-tech" industries may still incorporate technology to enable them to achieve higher productivity levels.

2.6.4.2 Systems of Innovation[7]

We alluded in Sect. 2.3.3.3 to the limitations associated with the linear model. In response to these limitations, attention turned in the late 1980s to innovation systems, which look at the process of technological advancement from a more systemic, interactive, and evolutionary standpoint. Innovation systems can be viewed from either the national, regional, or sectoral level.

Freeman (1987) defined innovation systems as "…the network of institutions in the public and private sectors whose activities and interactions initiate, import, modify, and diffuse new technologies." This approach focuses on the linkages among the agents involved in innovation, including private enterprises, universities, and public research institutes. Technical advancement and innovation are viewed as the outcome of relationships among these entities targeting the production, dissemination, and application of knowledge. The innovative performance of a country, according to this view, depends on how these actors relate to each other as elements of a collective system of knowledge creation and use.

For policymakers, an understanding of the national innovation system, or NIS, can help identify leverage points for enhancing innovative performance and overall competitiveness. It can help pinpoint mismatches within the system, both among institutions and with respect to government policies that hinder technology development and innovation; as well as the absence of necessary institutions within the system. Policies that seek to enhance the innovative capacity of firms, particularly their ability to identify and absorb technologies, are most valuable in this context. (OECD 1996; Lundvall et al. 2007; Edquist 2005).

Regional innovation systems (RIS) are closely related to clusters. An RIS features (1) firms that interact with each other, in particular those that tend to cluster; and (2) supporting regional infrastructure including private research laboratories, universities and colleges, and technology transfer agencies. An RIS can contain several clusters as long as there are firms and knowledge organizations that interact systematically within the boundaries of that region. Policymakers at sub-national levels may then focus on networking among firms and development of common infrastructure to support these systems.

An additional innovation systems concept is that of sectoral systems of innovation. Sectoral systems of innovation (Malerba 2004) stress the nature, structure, organization, and dynamics of innovation and production at the sector level; and focus primarily on firms, capabilities, and learning processes as the major drivers of innovation and growth, while also emphasizing the role of other actors in the system, including individuals, users, universities, government, and financiers and the linkages among them.

---

[7] This section draws substantially on Chap. 4 of Vonortas and Aridi (2012).

The innovation systems concept, although criticized for its lack of detail, is important because of the changing dynamics of international competition. Historically, advanced countries could assume that they would appropriate the benefits of technologies they produced. However, the rapidly developing capabilities of non-Western countries, particularly in Asia, to adopt and diffuse technologies produced elsewhere implies a need for a systemic approach to capturing those benefits; one embodied in the systems approach to innovation. This implies an important coordination role for policymakers to assure that the elements of the innovation system are working effectively together.

References

Abramowitz, M. (1986). Catching up, forging ahead, and falling behind. *Journal of Economic History, 46*(2), pp. 386–406.

Acs, Z., & Audretsch, D. B. (2001). Innovation in large and small firms: An empirical analysis. *The American economic review, 78*(4), 678–690.

Acs, Z., et al. (2013). National systems of entrepreneurship: Measurement systems and policy implications. George mason university school of public policy research paper no. 2012–08, 5 August 2013.

Albu, N. (2011). Research and development spending in the EU: 2020 growth strategy in perspective. Working paper FG1 no. 8, December 2011, SWP Berlin.

Auerswald, P., & Branscomb, L. (2003). Start-ups and spin-offs: Collective entrepreneurship between invention and innovation. In David H. (Ed.), *The emergence of entrepreneurship policy: Governance, start-ups, and growth in the knowledge economy*. Cambridge: Cambridge University Press.

Baumol, W. J. (1968). Entrepreneurship in economic theory. *American Economic Review Papers and Proceedings, 58*:64–71.

Baumol, W. J., Blackman, S. A. B., & Edward N. W. (1989). *Productivity and American leadership the long view*. Cambridge: MIT Press, c1989.

Cohen, W. M., & Levinthal, D. A. (1989). Innovation and learning: The two faces of R and D. *The Economic Journal, 99*, 569–596.

Cohen, W. M. (2010) Chapter 4: Fifty years of empirical studies of innovative activity and performance. In Hall and Rosenberg (Eds.), *Handbook of the Economics of Innovation*, Vol. 2, pp. 182–183.

Dunning, J. (1993). *Multinational enterprises and the global economy*. Harlow: Addison-Wesley.

Dutta, S., & Lanvin, B. (2013). High level segment: 1st July 2013. *The global innovation index 2013: The local dynamics of innovation*. http://www.globalinnovationindex.org/userfiles/file/blog/blog-july-1.pdf.

Edquist, C. (2005). Systems of innovation: perspectives and challenges. In J. Fagerberg, D. Mowery, & R. R. Nelson (Eds.), *The Oxford Handbook of Innovation*. Norfolk: Oxford University Press.

EURAB (European Advisory Research Board). (2004). SMEs and ERA, report and recommendations, EURAB 04.028-final. http://ec.europa.eu/research/eurab/pdf/eurab_04_028_sme_era.pdf.

Freeman, C. (1987). *Technology policy and economic performance: Lessons from Japan*. London: Pinter.

Greenhalgh, C. & Rogers, M. (2010). Chapter 9: Models of economic growth. *Innovation, intellectual property, and economic growth* (pp. 213–242). Princeton: Princeton University Press.

Hall, B. H. (2005). Innovation and diffusion. In Fagerberg, J., Mowery, D. C., & Nelson, R. R. (Eds.), *The Oxford Handbook of Innovation*. Oxford University Press.

Jacobs, J. (1969). *The economy of cities*. New York: Random House.

Lucas, R. (1988). On the mechanics of economic development. *Journal of Monetary Economics, 22*, 3–42.

Lundvall, et al. (2007). National systems of production, innovation, and competence-building. In Polenske, K. R. (Ed.), *The Economic Geography of Innovation*. Cambridge University Press.

Malerba, F. (2004). Sectoral systems of innovation: Basic concepts. In F. Malerba (Ed.), *Sectoral systems of innovation*. Cambridge University Press.

Morris, R. (2011). 2011 high-impact entrepreneurship global report. Endeavor center for high-impact entrepreneurship and the global entrepreneurship monitor.

OECD. (1996). The knowledge-based economy. Technical report. Organization for economic cooperation and development.

Porter, M. E. (1990/1998). *The competitive advantage of nations*. New York: Free Press.

Rodrik, D., & Hausmann, R. (2002). Economic development as self-discovery. NBER Working Paper Series #8952.

Rosegger, G. (1986). The Economics of Production and Innovation: An Industrial Perspective. Pergamon Press.

Romer, P. (1986). Increasing returns and long run growth. *Journal of Political Economy, 94* (2):1002–1038.

Schwab, K. (2013). The global competitiveness report 2012–2013. World economic forum. http://www3.weforum.org/docs/WEF_GlobalCompetitivenessReport_2012-13.pdf.

Schumpeter, J. (1942). Capitalism, Socialism, and Democracy. Routledge.

Solow, R. M. (1957). Technical change and the aggregate production function. *The Review of Economics and Statistics, 39*(3), 312–320.

Steen, J. V. (2012). Modes of public funding of research and development: Towards internationally comparable indicators. OECD Science, technology and industry working papers, 2012/04, OECD Publishing. http://dx.doi.org/10.1787/5k98ssns1gzs-en.

Stokes, D. E. (1997). *Pasteur's quadrant: Basic science and technological innovation*. Washington, DC: Brookings Institution Press.

Stoneman, P., & Battisti, G. (2010). Chapter 17: The diffusion of new technology. In Hall & Rosenberg (Eds.), *Handbook of the Economics of Innovation*, Vol. 2, pp. 733–760.

Verspagen, B. (2005). Innovation and economic growth. In J. Fagerberg, D. C. Mowery, & R. R. Nelson (Eds.), *The Oxford handbook of innovation*. Oxford: Oxford University Press.

Vonortas, N. S. (2013). Justifying Public Intervention in Research and Development.

Vonortas, N., & Aridi, A. (2012). The Innovation Policy Handbook. Center for International Science and Technology Policy, George Washington University, 2 June 2012.

Wikipedia: The Free Encyclopedia. (2014). Everett Rogers. http://en.wikipedia.org/wiki/File:Diffusionofideas.PNG.

Chapter 3
University Entrepreneurship: A Survey of US Experience

Cherilyn E. Pascoe and Nicholas S. Vonortas

3.1 Introduction

The rise in the commercialization of academic research achieved in the United States has sparked a heated debate to determine the components in the fuel of this success. As U.S. universities expand their patenting, licensing, and commercializing of research, their potential to drive domestic innovation and economic growth increases. However, there is a balancing act to be achieved: creating new innovations while not decreasing the university's primary role of education, research, and community outreach (Smith et al. 2010).

From an international comparative perspective, American universities have been uniquely successful in the intensity of their contributions to entrepreneurship (Rosenberg and Hart 2003). While research universities represent only a small portion of higher-education institutions, they are essential to economic success (Atkinson and Pelfrey 2010). The core activities of research performed by universities serve to investigate fundamental problems and create new knowledge, yet research also serves as an engine of innovation for the benefit of industry. New products and new industries, as well as trained graduates for new jobs, rely on the diffusion of knowledge that stems from academic basic, long-term, and applied research. While the spillover of academic research to enterprises varies depending on industry, many industries have relied on university research for real-world commercial purposes in agriculture, aerospace, biotechnology, medicine, software, computers,

C. E. Pascoe (✉)
Center for International Science and Technology Policy, The George Washington University, Washington, D.C., USA
e-mail: cherilyn.pascoe@gmail.com

N. S. Vonortas
Center for International Science and Technology Policy & Department of Economics, The George Washington University, Washington, D.C., USA
e-mail: vonortas@gwu.edu

© The Editor(s) 2015
N. S. Vonortas et al. (eds.), *Innovation Policy,* SpringerBriefs in Entrepreneurship and Innovation, DOI 10.1007/978-1-4939-2233-8_3

telecommunications, as well as social sciences industries such as network systems and communications, financial services, and transportation and logistical services (NRC 2010; NRC 2003). It is often argued that the American commercial success in high-technology sectors of the economy owes "an enormous debt to the entrepreneurial activities of American universities" (Rosenberg and Hart 2003).

Federal funding of university research has resulted in numerous and important commercial applications. For example, consider the list of the 50 most important innovations and discoveries funded by the National Science Foundation in its first fifty years. The *Nifty50* list includes innovations that have been commercialized from NSF university funding, most of which have applications in the high-technology sector, including, barcodes, fiber optics, the MRI, CAD/CAM software, speech recognition technology, data compression technology, and the Internet (NSF 2000).

A recent Information Technology and Innovation Foundation report found that universities and federal laboratories have become more important sources of the top 100 innovations over the last 35 years. In 1975, industry accounted for more than 70% of the 100 most significant R&D advances; by 2006, academia was responsible for more than 70% of the top 100 innovations (Mitchell 2010).

Indeed, university research plays a role in not only new products, but also the creation of new industries. For example, the Internet began as university research in the 1960s and 1970s as the ARPANET and Aloha network, then explored in industry research as Xerox's PARC Universal Packet (Pup) in the 1970s, eventually leading to a billion dollar industry in DECnet and TCP/IP networking, which in turn enabled numerous other innovations (NRC 2003b).

The creation of new technologies and industries rests partially on the transfer of new knowledge to industry through support of academic research and the movement of scientific talent out to the private sector, in the form of trained graduates. Such transfer is the process by which science and technology knowledge is diffused throughout human activity, flowing "to those who can apply it to practical problems in Government, in industry, or elsewhere" (Bush 1945). The transfer can occur through either the transfer of basic scientific knowledge into technology or the adaptation of existing technology to a new use (Brooks 1966). The formal concept of technology transfer is said to have originated from Vannevar Bush, in a letter to President Roosevelt following the Second World War, who recognized not only the value of university research to national defense, but also how research benefits industry, by increasing the flow of new knowledge to industry through the support of academic science (Bush 1945; COGR 1999).

University technology transfer takes many forms. The National Academies (2010) outlines the variety of technology transfer mechanisms: (1) trained graduates: movement of students from training to private employment; (2) publication of research results in literature; (3) personal interaction (professional meetings, conferences, seminars, industrial programs); (4) firm-sponsored university research projects; (5) multi-firm arrangements such as university-industry research centers; (6) individual faculty or student consulting with individual private firms; (7) entrepreneurial activity of faculty and students occurring outside university; (8) licensing of intellectual property to established firms or new start-up companies (NRC 2003,

2010). It is important to realize that technology transfer occurs in both directions. These forms of technology transfer allow mutually beneficial relationships in which research findings and business information can be shared between and amongst universities, the government, and the private sector.

Only recently, the volume of research on the topic of the entrepreneurial university has been increasing, seeming to correspond with increasing levels of entrepreneurship in universities around the world. Rothaermel et al. (2007), in taxonomy of the literature from 1982 to 2005, observe that most of the journal articles have only been published since 2000. While a small number of universities were taking an active role in innovation, most academic scientists and research universities abstained from commercializing research until recently (Etzkowitz 2003). Basic concepts such as "invention" and "innovate" were highly contested and unfamiliar, even on the most entrepreneurial university campuses. The increase in university entrepreneurship can be attributed to the industry's growing demand for technological innovation and universities' search for new sources of funding, resulting from reduced federal funding for research (Thursby and Thursby 2002; Rothaermel et al. 2007). The formation of new actors, relationships, and networks in academia and industry highlight the tension in the transition from the research university to the entrepreneurial university (Rhoten and Powell 2007).

Etzkowitz (2003) argues that the entrepreneurial university encompasses and extends the research university; it includes the primary missions of teaching and research, adding a new mission of economic and social development. This chapter is focused on a discussion and critiques of three main areas: (i) the Bayh-Dole Act which redefines the landscape to allow universities to keep patent rights stemming from research; (ii) the university, including organizational and cultural issues stemming from pressures for universities to take an active role in innovation; and (iii) the technology transfer office, the focal point for how universities interact with commercial interests.

3.2 Bayh-Dole Act

The Bayh-Dole Act was intended to facilitate U.S. technological innovation by standardizing the intellectual property ownership of inventions created with federally funded research. In 1980, the federal government expenditures for research and development totaled $ 55.5 billion (in constant 2000 US dollars). The federal money, appropriated through the United States Congress, is primarily used to support research and development to meet the mission requirements of federal agencies and departments as well as support basic research not being performed in the private sector. Generally the government retained title to inventions made with federal funds regardless of who performed the research (usually by universities and federal laboratories) and negotiations to use such government patents resulted in primarily non-exclusive licenses. Yet the agencies and departments had 26 different policies regarding the use of federally funded research (Schacht 2011). The

Patent and Trademark Law Amendments Act of 1980, commonly referred to as the Bayh-Dole Act after its two main sponsors former United States Senators Robert Dole and Birch Bayh, was meant to replace the bureaucratic red tape with a single national policy. The Act allows the inventor—universities or other non-profit institutions—to retain intellectual property ownership from federally sponsored research and development (Schacht 2009). The Bayh-Dole Act was intended to minimize the likelihood that government funded inventions would languish in the absence of incentives to license and develop the invention (Mowery and Sampat 2001b; Schacht 2011).

The Bayh-Dole Act has been seen as successful in promoting expanded commercialization of the efforts of federally funded research. The *Economist* writes, "possibly the most inspired piece of legislation to be enacted in America over the past half-century was the Bayh-Dole act of 1980 … more than anything, this single policy measure helped reverse America's precipitous slide into industrial irrelevance" (Innovation's Golden Goose 2002). Several university and education associations agree that "the current legal framework for university technology commercialization, as set forth by the Bay-Dole Act of 1980 and implementing regulations is effective, and needs to be maintained" (Smith et al. 2010). Prior to 1981, U.S. universities issued fewer than 250 patents annually and discoveries were infrequently commercialized (Dickinson 2000). According to a recent U.S. licensing activity survey by the Association of University Technology Managers (AUTM), a non-profit association of academic technology transfer professionals, in 2010, 4469 U.S. patents were issued to U.S. universities, 651 new companies were formed, 657 new products were introduced, and over 3600 startup companies were still operating based on university inventions (AUTM 2011). According to a Biotechnology Industry Association study, between 1996 and 2007, as much as $ 187 billion of U.S. gross domestic product, resulting in 279,000 jobs can be attributed to university technology licensing (Smith et al. 2010; Roessner et al. 2009).

Bernadine Heal, former National Institute of Health director, credited the Bayh-Dole Act for the development of the entire biotechnology sector (Dickinson 2000). The pursuit of patenting at universities has expanded because of university research. Most federally sponsored research is dedicated to health-related research, with the National Institute of Health investing approximately $ 30 billion in research, annually (NIH 2012). Examples of biotechnology innovations resulting from university research include an artificial lung surfactant for babies (University of California), a treatment for Crohn's disease (Washington University in St. Louis), non-toxic therapies for Chagas disease (Washington University and Yale University), haemophilus B conjugate vaccine (University of Rochester), recombinant DNA technology (Stanford University and University of California) (BayhDole 2006). This list represents only a small number of all innovations that have resulted from university research. Although the Bayh-Dole Act alone may not have created the biotechnology revolution, the incentives it provides for technology transfer are a critical institutional factor in innovation. In addition, there are many reasons for this growth in commercialization stemming from the passage of the Bayh-Dole Act: universities have substantially increased investment in technology transfer programs, faculty have become aware of the commercial potential of their research results, and industry has realized the benefits of collaborating with universities.

While much of the expanded commercialization has been attributed to the passage of the Bayh-Dole Act, there is little evidence and much discussion critique surrounding the argument (Mowery et al. 2004). Mowery and Sampat (2005) discussed that the effects of Bayh-Dole, including institutional responsibilities and incentives for patenting and licensing, have led to more universities entering into technology transfer activities and establishing technology transfer offices than without the law. Shane (2004), after analyzing patents across 117 lines of business, found that the Bayh-Dole Act provided incentives for universities to increase patenting in fields where licensing is effective in producing new technical knowledge. In contrast, Coupe (2003) analyzed National Science Foundation research expenditure data and university patents and found that establishing Technology Transfer Offices, rather than the Bayh-Dole Act, had a positive effect on patenting activities of universities. Mowery et al. (2001) found that the Bayh-Dole is only one of several factors, in addition to federal support and portfolio of research, behind the rise in university patenting and licensing. In addition, patenting and university involvement in patents was occurring prior to the Bayh-Dole Act (Mowery and Sampat 2001b). Indeed, it was the universities who were active in technology transfer that lobbied for the passage of the law (Etzkowitz 2003).

The passage of the Bayh-Dole Act was the result of years of opposition and emotional debate. Specifically, United States Senator Long was concerned that taxpayers would not receive a direct benefit of government-funded research (Baumel 2009). His opposition led to the inclusion of compromises, which, as explained by the preamble of the law, "ensure that the Government obtains sufficient rights" and "protect the public against nonuse or unreasonable use of inventions". The government's "march-in" right is one of the most contentious provisions added to Bayh-Dole. This provision allows federal agencies to assert march-in rights if the holder of the patent fails to commercialize the invention for the general public or if action is necessary to protect safety or health needs. Though in theory the provision is quite powerful, and "could be wielded wisely to greatly benefit the public investors in federally-funded inventions" (Ritchie de Larena 2007), to date, no federal agency has asserted its march-in rights.

Thirty years after the passage of the Bayh-Dole Act, there have been critical reviews of its problems, discussion of its successes, and opportunities to update the law. Proponents of the Act say that "the success of the Bayh-Dole Act speaks for itself" (Baumel 2009), with evidence from the substantial amount of innovations, particularly in the biotechnology sector, that have been developed due to licensing of university research over the last three decades. In general, the Bayh-Dole Act has served its purpose as a legal framework for technology commercialization. It created a stable, regulated environment for the arrangement of intellectual property rights proceeding from federally funded research activities (Etzkowitz 2003). It may not be the primary reason for university patenting, but it did provide a framework and hastened many universities entry into patenting and licensing activities (Mowery and Sampat 2001b). Indeed the visibility of the success of the Bayh-Dole Act has raised some concerns associated with challenges to university's primary role in research. Some of the criticism of the Bayh-Dole Act stems from the question of whether a university should be engaged in entrepreneurial activities. For example, the American Association of Universities and other education associations share the concern Congress raised in the Act that commercialization might encumber the university's primary education, research, and discovery missions.

3.3   The Entrepreneurial University

In 1981, William Massey, the University of Stanford's vice president for business and finance, told the university's faculty that "it's important that we find means by which the university can participate in the entrepreneurial returns that come from those things that we create here"; he also was worried that others outside the university were receiving "windfall profits" from the "natural income of the fruits of research" (quoted in Washburn 2005). Perhaps, one of the most pronounced conflicts surrounding a university's governance is to support entrepreneurial activities without losing control of its primary education, research, and public service missions. Most agree with Massey that closer cooperation among industry and academia can augment funding and revenue sources, increase technology transfer and stimulate innovation. However, such collaboration and entrepreneurial activity at universities raises questions about the impact on traditional academic procedures, providing for secrecy in research, diversion of research priorities to applied rather than basic research, and opportunities for conflict of interest (Schacht 2009; NRC 2010).

3.3.1   Incentive Structures

While the motivation to develop innovative technologies is clear and several opportunities exist, most universities have not changed core activities or associated incentive systems (Brouwer 2005). Most studies recommend providing incentives and rewards directly to faculty to encourage invention disclosures and commercialization activities (Henrekson and Rosenberg 2001; Jensen and Thursby 2001; Friedman and Silberman 2003; Debackere and Veugelers 2005). There are a number of reasons that the potential values of incentive systems are important to faculty. First, there is the embryonic nature of these technologies, as over 70 % of technologies require inventor involvement in the technology transfer from university laboratory to industry—even after a license agreement is signed (Thursby et al. 2001; Coupe 2003; Jensen and Thursby 2001). Second, very few universities allow faculty to own their inventions, regardless of whether the research funding came from industry or the government (Thursby et al. 2001). Finally, scientists are more likely to face pressures to focus on basic research and publish their results in journals, rather than patent or license an invention. Therefore, there are a number of reasons why scientists, with proper incentives, would be more likely to coordinate with industry to disclose their inventions and be involved in technology transfers.

Many scholars find that incentives do not motivate faculty to engage in entrepreneurial activities. Colyvas et al. (2002), by analyzing technology transfer processes from Columbia and Stanford Universities, reported that financial incentives play little role in motivating faculty to embark on invention-producing research projects. Friedman and Silberman (2003), using data from AUTM, found that there is a weak dependence between rewards to inventors and the number of licenses executed and number of licenses generating income. Markman et al. (2004) found a negative correlation between monetary incentives given to scientists whose inventions were licensed and the number of equity licenses and the number of start-up companies. In general, universities not only have struggled to provide proper incentives to faculty to commercialize research, but there is also a conflict as to whether incentives are necessary to encourage an entrepreneurial culture within the university.

3.3.2 Concern About Publication Delay and Increased Secrecy

Former Stanford University president, Donald Kennedy observed: "To those who had worried about technology transfer, it's a huge success. To others, who expressed concern about university/corporation relations or mourn the enclosure of the scientific 'knowledge commons', it looks much more like a bad deal" (Rhoten and Powell 2007). Delays in publication and free flow of information from universities are serious factors for impeding innovation. Professor Richard Florida said it "may well discourage or even impede the advancement of knowledge, which retards the efficient pursuit of scientific progress, in turn slowing innovation in industry" (Schacht 2009).

There is the worry that withholding and secrecy are more common among the most productive and entrepreneurial faculty. A series of studies in the 1990s by Blumenthal, Louis, and colleagues at the Harvard Medical School called attention to delays in publication of biomedical research results (Blumenthal et al. 1997; Campbell et al. 2002; Louis et al. 2001). Specifically, they indicated that approximately 20 % of life sciences researchers delayed publication for reasons associated with commercialization considerations, such as obtaining intellectual property protection before disclosing results (Blumenthal et al. 1997). Louis et al. (2001) discovered that life sciences faculty involved in entrepreneurial behavior are more likely to be secretive about their research, but are not any less productive in their faculty roles. Huang and Murray (2009) found that while there are 4000 human gene patents, there is a measurable decrease in the amount of published genetic knowledge in journals. Within the field of medicine and genetics, commercial consideration and the broader patent landscape play an important role in withholding research results and delaying publication (Huang and Murray 2009). In more recent years, worry about publication delays and secrecy has diminished. As more university researchers are filing provisional patent applications in advance of formal applications, the incentive to postpone disclosure or delay publication has been reduced (NRC 2010).

3.3.3   Diversion of Research Priorities

James Severson, president of the Cornell Research Foundation, testified before the United States House of Representatives Committee on Judiciary, that the Bayh-Dole Act encourages research that is attractive to faculty (Severson 2000). Yet because of the profit motive to commercialize research, there is the question of whether faculty involvement in entrepreneurial activities diverts research priorities away from fundamental scientific exploration toward work on applied research with practical applications (NRC 2010).

Some scholars find that the entrepreneurial university drives more applied and problem-solving research, moving away from basic research, thus aggravating the conflict between advancing knowledge and generating revenues (Lee 1996; Powell and Owen-Smith 1998). A majority of scholars, however, find no evidence that the shift towards applied research occurs at the expense of basic research (Van Looy et al. 2004). In fact, data collected by the National Science Foundation show that the split between basic and applied research expenditures has not changed. In 1980, basic research comprised 66% of all academic research expenditure, while applied research comprised 33% of the total. In 2007, the percentage of academic research expenditures devoted to basic research actually increased to 76%, while applied research declined to 24% of the total (Schacht 2009). Not only has there been an increase in the percentage of academic basic research compared to applied, but Thursby et al. (2007) found that commercial activity increases the level of all research efforts. Similarly, Van Looy et al. found that entrepreneurial activity increases publication outputs without affecting the nature of the publications. They thus found not only that entrepreneurial activity has increased research efforts but that entrepreneurial performance and scientific performance do not hamper each other (Van Looy et al. 2004).

3.3.4   Conclusions

There are a number of positive effects of entrepreneurial activities at a university. First, participation in networking increases along with formal technology transfer. Also, income from new technologies has increased research funding available to departments and faculty (NRC 2010). Overall, despite concerns that increased commercialization comes at the expense of the universities' primary role of education, research, and community outreach, there is little research to support these assertions. Commercially oriented faculties are not less likely to publish in academic journals (in fact, entrepreneurial performance increases publication outputs) (Van Looy et al. 2004; Azoulay et al. 2009). Commercial incentives have not shifted effort away from fundamental research questions (Thursby et al. 2007); in fact, basic research has increased compared to applied research (Schacht 2009). In addition, institutional concerns to protect intellectual property no longer result in delays in publication of research results (NRC 2010).

Scholars have attempted to reconcile opposing issues in university commercialization by concluding that the mission of universities requires both traditional (teaching and research) with entrepreneurial roles (economic and social mission) (Etzkowitz 2003). While the actions, incentives, and organizational culture for university administrators, scientists, and firms differ greatly, the two cultural roles for universities (scientific and entrepreneurial) may actually complement and reinforce each other. Siegel et al. (2004) argued that greater faculty involvement, greater resources for the technology transfer office, and a mutual understanding in terms of culture will benefit university innovation technology transfer.

3.4 Technology Transfer Offices

The first documented success story in university licensing is from the University of Wisconsin-Madison. The Wisconsin Alumni Research Foundation was formed in 1925, as a nonprofit foundation charged with administering patents and licenses resulting from faculty research. In 1924, Professor Harry Steenbock published and patented his research demonstrating that vitamin D could be activated in food. After the patent application, Quaker Oats Company offered Steenbock $ 900,000 (approximately $ 11.9 million in 2012 dollars) for exclusive rights to the patent. Steenbock rejected the offer and assigned patent rights to the Wisconsin Alumni Research Foundation. The Foundation paid Steenbock 15 % of the net income from the patent in order to provide faculty with an incentive to patent. Since making its first grant of $ 1200 in 1928, the Wisconsin Alumni Research Foundation has contributed more than $ 1 billion to University of Wisconsin-Madison, including funding for research, facilities, land and equipment, and faculty and graduate student fellowships (Wisconsin Alumni Research Foundation 2010).

When the Bayh-Dole Act was passed in 1980, there were only 25 technology transfer offices in the United States. By the twenty-fifth anniversary of the Act, in 2005, there were 3300 such offices (Ritchie de Larena 2007). An Association of University Technology Managers (AUTM) figure shows the growth of university TTOs over the years. The purpose of the technology transfer office (TTO) is to promote the utilization of inventions from university research. It allows universities and researchers to capitalize on the rights they gain through the Bayh-Dole Act while attempting to allay concerns regarding conflict of interest. Rather than relying on researchers to commercialize their inventions or implementing broad innovation strategies, many universities have channeled their innovation activities through a centralized TTO. TTOs are dedicated to identifying research that has potential commercial interest, providing legal and commercialization support to researchers, assisting with questions relating to marketability and funding sources, and serving as a liaison to industry partners, interested in commercializing university technologies (Fig. 3.1).

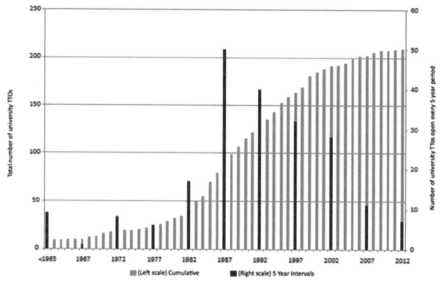

Fig. 3.1 Growth of university TTOs. (Source: AUTM 2013)

3.4.1 Effectiveness in Commercializing University Research Results

The US National Academies conclude that universities should evaluate their individual TTO against their own missions "and yet recognize that they are part of a larger education and research enterprise" (NRC 2010). The effectiveness of a TTO is typically measured by its commercial output, including licensing (number of licenses, licensing revenue), equity positions, coordination capacity (number of shared clients), information processing capacity (invention disclosures, sponsored research), and royalties and patents (number of patents, efficiency in generating new patents) (Rothaermel et al. 2007).

There is a wide range of success with inventions and licensing occurring at institutions across the country. According to a 2007 Association of University Technology Managers (AUTM) survey, the license income for select institutions ranged from $ 0 to almost $ 800 million with the total license income reported for the 194 institutions at $ 2.7 billion (House of Representatives Committee on Science and Technology 2010). When Kordal and Guice (2008) grouped institutions according to size, they still found large differences in revenue, invention disclosure, patenting rates, licensing, and start-up companies. On the one hand, it would be useful to know to the extent disparities among universities reflect differences in the organizational structure of TTOs, including staffing, funding sources, relationship with faculty, and procedures for sharing royalties with the inventor and department. On the other hand, disparities could reflect differences in structural factors of the university such as scale and specialization of research portfolios, public vs. private status, historical reputation, and geographical proximity to potential investors and industry partners (NRC 2010).

The disparities in revenue generation from TTOs among universities are very significant. While many universities rushed to open TTOs after the Bayh-Dole Act, only a few raised income from licensing their patents. In 2012, the top 5% of earners, 8 universities, took 50% of the total licensing income of the university system. The top 10%, 16 universities, took nearly three-quarters of the system's income. The figure below shows the wide asymmetric distribution of licensing gross income. Moreover, only 37 universities have been able to reach the top 20 of licensing revenue any given year over the last ten years creating an elite club of universities with stable membership (Valdivia 2013).

3.4.2 Performance in Launching New Firms

The attractions of using university-developed inventions to create new start-up companies (a new company created to commercialize a particular technology) have become widely recognized. A study by the Kauffman Foundation found that net job growth occurs in the U.S. economy only through startup firms. Between 1977 and 2005, existing firms are net job destroyers, losing 1 million jobs net combined per year; on the other hand, new firms add an average of 3 million jobs (Kane 2010). In addition, founding a company like Google or Yahoo! that becomes a global leader in the industry holds promise of financial benefits and brightens the reputation of the university. However, most faculty invention disclosures lend themselves to licensing agreements with existing companies rather than the formation of a start-up firm. According to an AUTM survey, the ratio of start-ups to licensing agreements with established firms ranges from 1:1.5–1:22 across institutions. However, sometimes commercialization of the invention is best suited via the creation of a start-up company and TTOs are beginning to place more emphasis on creating new business start-ups as an optimal commercialization path. As evidence of this increase, the number of start-up firms for commercialization of university research grew from 241 in 1994 to 555 in 2007 (NRC 2010).

There are conflicting ideas on the role TTOs should play in promoting the launch of new firms, ranging from no role in start-ups to a very involved role in helping start-up firms succeed. Some venture capital and angel investors believe that the university should play no constructive role in forming new start-ups other than licensing any underlying intellectual property. Some universities rely on TTOs to network ideas to early-stage investors. Some universities move beyond the capacities and resources of a TTO and have an innovation center independent of the TTO to help attract investors or seed capital funds from alumni contributors. Some universities even have access to incubators and science parks where their start-up companies can share low-cost space and services to help them succeed (NRC 2010). Overall, Di-Gregorio and Shane (2003) reported no effect on the start-up rates from local venture capital activity, presence of university-incubator, or whether the university is permitted to actively make venture capital investments in licensees (Fig. 3.2 and Table 3.1).

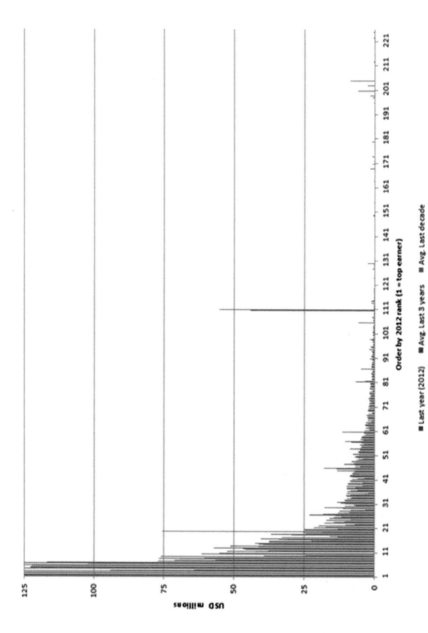

Fig. 3.2 Distribution of licensing gross income by university. (Source: AUTM 2013; cited in Valdivia 2013)

Table 3.1 Top earners of licensing gross income 2003–2012. (Source: AUTM 2013; cited in Valdivia 2013)

University	Rank 2012	Times in top 20 over the last decade
New York University	1	10
Columbia University	2	6
Massachusetts Institute of Technology	3	10
Princeton University	4	2
Northwestern University	5	7
Univ. of California System	6	10
University of Washington	7	10
Stanford University	8	9
Mount Sinai School of Medicine	9	7
University of Texas System	10	4
University of Massachusetts All Campuses	11	10
University of Minnesota-Twin Cities	12	9
University of Wisconsin-Madison	13	10
University of Rochester	14	10
University of Utah	15	8
University of Florida	16	10
University of Colorado System	17	5
California Institute of Technology	18	5
Emory University	19	6
Duke University	20	4
University of Illinois at Urbana-Champaign	21	1
University of Pennsylvania	23	1
University of Michigan-Ann Arbor	26	6
Harvard University	29	5
State University of New York System	31	3
Iowa State University	34	1
University of Nebraska – Lincoln	38	1
University of Georgia	44	4
University of Iowa	45	6
Washington University in St. Louis	53	1
Michigan State University	60	3
University of South Florida	81	1
Florida State University	86	2
Wayne State University	105	2
Wake Forest University	110	9
Eastern Virginia Medical School	129	1
University of Texas Southwestern Med Center	150	1

3.4.3 Relations with Private Research Sponsors

Collaboration provides many benefits to firms, including increased access to new university research, knowledge, and discoveries. Benefits for faculty members include additional insights into their research and funds for graduate students and laboratory resources (Lee 2000). However, some argue that the strong property orientation ushered in by the Bayh-Dole Act has prevented productive university-industry relations (Rhoten and Powell 2007). Johnson (2007), Vice President of Hewlett-Packard, discussed how aggressive university patenting and overvaluing of intellectual assets impedes university-industry collaboration, encouraging companies to find other research partners. Kramer (2008) found that some firms prefer foreign university partnerships because academic institutions abroad are less insistent on intellectual property ownership and complex agreements. Negotiation challenges between universities and industry were also reported by Hertzfeld et al. (2006). Several scholars, including Thursby and Thursby (2006) have discussed that universities play a role in the decision to offshore an industry's research and development activities; however, they show that other determinants such as research cost and skilled talent are much more of a factor in the decision than attitudes towards academic institutions (NRC 2010). The National Council of University Research Administrators and the Industrial Research Institute offer a solution to the contentious issues in negotiation. They recommend that universities avoid licensing future inventions, instead focusing on a general framework for future intellectual property, but not a specific agreement. In addition, research needs vary by industry. Jon Soderstrom, AUTM president, wrote that most universities are not adapted to the needs of the information technology industry and other physical sciences technology, due to their focus on commercialization of biotechnology (Kramer 2008).

3.4.4 Critiques of Technology Transfer Offices

The TTO is intended to facilitate the transition between academic research and commercialization. While some universities' TTOs are effective in disseminating inventions, others have become hindrances to technology transfer with levels of administration and bureaucracy (Litan et al. 2007). The generation of TTOs that followed Bayh-Dole was not a goal of the legislation, but rather one of its byproducts, or what Litan et al. (2007) called an "unintended consequence of the Act". Leading criticisms for TTOs include: (1) university administrators have stronger incentives to use TTOs as generators of revenue rather than focusing on transferring technologies, neglecting some inventions with little profit potential (Kenney and Patton 2009); (2) academic red tape with some TTO's tying up university faculty with patent-related paperwork that detracts from the research mission; (3) TTO personnel have little knowledge of the research results, invention, and marketplace for the invention; (4) intellectual property management and technology transfer revenue distribution policies are confusing and rigid (NRC 2010; Ritchie de Larena 2007).

Efforts to bridge the gaps in university-industry relationships have been moderately productive (NRC 2010).

3.4.5 Alternative Approaches to TTO Commercialization Activities

In general, there has been little experimentation and universities have been slow to move to more flexible structures for managing intellectual property. One possible idea to bridge the gaps in technology transfer, discussed by Litan and Mitchell (2010), is to create an open, competitive licensing system for university technology. Currently, a university professor with an invention may commercialize it only by using the university's technology licensing office. Litan and Mitchell argued that this is an inefficient monopolistic arrangement which slows the process of commercialization. "We have a massive bottleneck of innovation on our campuses. Even though federal [research] funding from the National Institutes of Health has more than doubled over the past 15 years, the number of new drug approvals has fallen from 40 to 50 a year down to 12," said Mitchell, at the Kauffman Foundation. "As the federal government dedicates billions of dollars in research funding to clean energy, we cannot let this pattern be repeated." An amendment to the Bayh-Dole Act would allow faculty members to choose their own licensing agents. They argued that this would increase competition and speed up commercialization, while still allowing the university to collect royalties on the invention. Conversely, the National Academies argued against breaking the status quo, saying that there is no evidence that inventors have knowledge and skills of management and licensing that is superior to the personnel of a TTO (NRC 2010).

Some feel that enhanced intellectual property rights at the early stages of research may hinder the dissemination of scientific knowledge. Sampat (2003) argued that patenting can promote innovation, but the effects, particularly on recently developing research efforts remains a concern, requiring further evidence to determine the net effects. As universities make intellectual property claims to fragmented knowledge and to early stage inventions, patents can increase the costs of research and slow innovation. In addition to the concern of monopolization and control from TTOs, some scholars emphasize a move towards open source and open access (Kenney and Patton 2009; Rhoten and Powell 2007). Kenney and Patton (2009) discussed an alternative to address the current "dysfunctional" arrangements in university technology dissemination. They want to make all inventions publically available through a public domain strategy or through a requirement that all inventions be licensed non-exclusively. From a societal standpoint, the rise of open innovation is salient; for example, Android, Wikipedia, YouTube are all examples of collaboration using open source strategies.

Open innovation would accelerate the adoption of new university-developed technologies. It seeks to co-develop and co-commercialize research and technology, regardless of where the research originated (Markman et al. 2008). In addition, placing inventions in the public domain would remove concerns that commercialization of research activities changes the university's culture and mission. This would solve the reconfiguring of U.S. science that has shifted from separate realms of university and commercial science to an increasingly interconnected world of public and proprietary science (Rhoten and Powell 2007).

Another possible approach for technology transfer is to create regional alliances, where universities establish consortia to improve mechanisms for commercialization. Universities would be able to share cost and expertise among the consortia. This approach is attractive to universities that have TTOs with limited budgets and limited expertise. For example, expertise is needed not only in technology transfer activities, such as patenting and licensing, but knowledge in all fields of the university's research portfolio and how industry may use that knowledge. However, regional consortia may face a number of conflicts, including coordination challenges or disputes over attribution (Litan et al. 2007).

Related to the regional alliance model, the Internet-based approach to technology transfer uses the Internet to facilitate commercialization of university inventions. This model uses matchmaking to match those who have ideas with those who are willing to commercialize them. An example of this approach is the iBridge Network, a Web-based platform by the Kauffman Foundation. Universities join the network and post information about their research activities. Faculty can share access to research tools, materials, and licensed technologies to acceleration innovation and lower transaction costs. Another example is www.Yet2.com which connects industry, academic institutions, public and non-profit organizations with a global network of scientists to managed intellectual property. It connects buyers and sellers of technologies (Markman et al. 2008). However, the success of Internet-based approaches to facilitate commercialization remains to be seen (Litan et al. 2007).

3.5   Conclusion

The appraisal of the influence of the Bayh-Dole Act, the university's entrepreneurship activities, and the success of technology transfer offices does not allow many unequivocal conclusions. The Bayh-Dole Act enabled universities and university faculty to take part in entrepreneurial activities and collaborate with industry. However, several scholars argue that the Act places profit at the center of government-funded research activities. Some argue that the university's entrepreneurship activities, by diverting priorities, delaying research publications, and incentivizing conflict of interest, conflict with its traditional role as a university. Other scholars show proof that this is not the case: they argue that, on the contrary, scientific and entrepreneurial activities enhance each other. Technology transfer offices were formed at universities to provide an effective way for universities and researchers to capitalize on the new rights they gain through Bayh-Dole, while providing an opportunity to allay concerns of conflict of interest. However, academic red tape and layers of bureaucracy at technology transfer offices have arguably become hindrances to effective technology transfer.

US-based research universities have been successful overall in creating new inventions for many high-tech industries, yet if universities fail to address their entrepreneurship role in the innovation system, economic growth could suffer. In a world in which economic activity is becoming more globalized and knowledge-intensive, it is important to rethink the university entrepreneurial structure in way that makes them remain a key player in the national innovation system and the economic environment. The transfer of new knowledge, through the support of academic research, plays an important role in economic growth through the creation of new products and industries. While some universities have been successful in mutually beneficial technology transfer for some time, the course of the entrepreneurial university as a widespread phenomenon has just begun. Policy decision makers argue that universities should take a broader view of their role in technology transfer and economic growth: entrepreneurship at the university ought to focus on maximizing the social impact of technology, rather the university's licensing revenue. For the Bayh-Dole Act to work as intended, universities must look beyond short-term profit and think about what is best for society as a whole (Lemley 2007).

Youtie and Shapira (2008) argue that the university has evolved from the traditional mission as a storehouse of knowledge to a knowledge factory. They add that in the future a third model for the university will emerge; the university will evolve to a knowledge hub that seeks to animate development, new capabilities, and innovation within the larger innovation system. The underlying shifts in the economy challenge universities to reorganize their research and education and reconsider ways in which they develop and disseminate knowledge.

References

Association of University Technology Managers (AUTM). (2011). U.S. Licensing Activity Survey: FY 2010.

Association of University Technology Managers (AUTM). (2013). U.S. Licensing Survey Activity FY 2013.

Atkinson, R. C., & Pelfrey, P. A. (2010). Science and the entrepreneurial university. *Issues in Science and Technology Policy, 26*(4), 39–48. http://www.issues.org/26.4/atkinson.html.

Azoulay, P., Ding, W., and Stuart, T. (2009). The impact of academic patenting on the rate, quality and direction of (public) research output. *The Journal Of Industrial Economics, 54*(4), 637–676.

Baumel, J. A. (2009). The Bayh-Dole Act: The technology revolution shows its age. Venture Capital Review. 22. Winter.

BayhDole. (2006). The Bayh-Dole Act at 25. BayhDole25, Inc. White Paper. April 17, 2006.

Bercovitz, J., Feldman, M., Feller, I., & Burton R. (2001). Organizational structure as a determinant of academic patent and licensing behavior: An exploratory study of Duke, Johns Hopkins, and Pennsylvania State Universities. *Journal of Technology Transfer, 26*(1–2), 21–35.

Blumenthal, D., Campbell, E.G., Anderson, M. S., Causino, N. A., & Karen Seashore L. (1997). Withholding of research results in academic life science: Evidence from a national survey of faculty. *Journal of American Medical Association, 277*(15), 1224–1228.

Brooks, H. (1966). National science policy and technology transfer. Proceedings of the Conference on Technology Transfer and Innovation. National Science Foundation. Washington, DC: US Government Printing Office.

Brouwer, M. (2005). Entrepreneurship and university licensing. *Journal of Technology Transfer, 30* (3), 263–270.

Bush, V. (1945). *Science the endless frontier: A report to the President by Vannevar Bush.* Washington, DC: United States Government Printing Office.

Campbell, E. G., Clarridge, B. R., Gokhale, M., Birnebaum L., Hilgartner S., Holtzman N. A., & Blumenthal, D. (2002). Data withholding in academic genetics: Evidence from a national survey. *Journal of American Medical Association, 287*(15), 1939–1940.

Colyvas, J., Crow, M., Gelins, A., Mazzoleni, R., Nelson, R. R., Rosenberg, N., & Bhaven, N. S. (2002). How do university inventions get into practice? *Management Science, 48,* 61–72.

Council on Governmental Relations (COGR). (1999). *The Bayh-Dole Act. A guide to the law and implementing regulations.* Washington, DC: Council on Governmental Relations.

Coupe, T. (2003). Science is golden: Academic R & D and university patents. *Journal of Technology Transfer, 28*(1), 31–46.

Debackere, K. and Veugelers, R. (2005). The role of academic technology transfer organizations in improving industry science links. Research Policy, *34*(3), 321–342.

Dickinson, T. (2000). Reconciling research and the patent system. *Issues in Science and* Technology, 16(4). http://www.issues.org/16.4/dickinson.htm. Accessed. 1 Mar 2012.

DiGregorio, D. D., & Shane, S. (2003). Why do some universities generate more start-ups than others? *Research Policy, 32,* 209–227.

Etzkowitz, H. (2003). Research groups as 'Quasi-Firms': The invention of the Entrepreneurial University. *Research Policy, 32,* 109–121.

Friedman, J., & Silberman, J. (2003). University technology transfer: Do incentives, management, and location matter? *Journal of Technology Transfer, 28*(1), 17–30.

Henrekson, M., & Rosenberg, N. (2001). Designing efficient institutions for science-based entrepreneurship: Lesson from the US and Sweden. *Journal of Technology Transfer, 26*(3), 207–231.

Hertzfeld, H. R., Link, A. N., & Vonortas, N. S. (2006). Intellectual property protection mechanisms in research partnerships. *Research Policy, 35*(6), 825–838.

House of Representatives Committee on Science, Space, and Technology. Science and Technology. (2010). From the lab bench to the marketplace: Improving technology transfer: Hearing charter. U.S. House of Representatives Committee on Science and Technology, Subcommittee on Research and Science Education. June 10, 2010.

Huang, K. G., & Murray, F. E. (2009). Does patent strategy shape the long-run supply of public knowledge? Evidence from human genetics. *Academy of Management Journal, 52*(6), 1193–1221.

Innovation's Golden Goose. (2002). *The Economist.* 14 December 2002, 3.

Jensen, R. A., & Thursby, M. C. (2001). Proofs and prototypes for sale: The licensing of university inventions. *American Economic Review, 91*(1), 240–259.

Johnson, W. C. 2007. Testimony on The Bayh-Dole Act (P.L. 96-517, Amendment to the Patent and trademark Act of 1980)—The Next 25 Years. July 17, 2007. Hearing to the U.S. House of Representatives Committee on Science and Technology. Subcommittee on Technology and Innovation.

Kane, T. (2010). Importance of startups job creation and job destruction. Kauffman Foundation Research Series: Firm Formation and Economic Growth. http://www.kauffman.org/what-we-do/research/firm-formation-and-growth-series/the-importance-of-startups-in-job-creation-and-job-destruction. Accessed 1 Mar 2012.

Kenney, M., & Patton, D. (2009). Reconsidering the Bayh-Dole Act and the current university invention ownership model. *Research Policy, 38*(9), 1407–1422.

Kordal, R., & Guice, L. K. (2008). Assessing technology transfer performance. *Research Management Review, 16*(1), 45–56.

Kramer, D. (2008). Universities and industry find roadblocks to R & D partnering. *Physics Today, 61,* 20–22.

Lee, Y. S. (1996). Technology transfer and the research university: A search for the boundaries of university-industry collaboration. *Research Policy, 25*(6), 843–863.

Lee, Y. S. (2000). The sustainability of university-industry research collaboration: An empirical assessment. *Journal of Technology Transfer, 25*(2), 111.

Lemley, M. A. (2007). Testimony on The Bayh-Dole Act—The Next 25 Years. July 17, 2007. U.S. House of Representatives Committee on Science and Technology. Subcommittee on Technology and Innovation.

Litan, R. E., Mitchell, L., & Reedy, E. J. (2007). The university as innovator: Bumps in the road. *Issues in Science and Technology, 23*(4), 57–66.

Litan, R. E., & Mitchell, L. (2010). Breakthrough Ideas for 2010. A faster path from lab to market. *Harvard Business Review, 88,* 52–53.

Louis, K. S., Jones, L. M., Anderson, M. S., Blumenthal, D., & Campbell, E.G. (2001). Entrepreneurship, secrecy, and productivity: A comparison of clinical and non-clinical life sciences faculty. *Journal of Technology Transfer, 26*(3), 233–245.

Markman, G. D., Gianiodis, P. T., Phan, P. H., & David, B. B. (2004). Entrepreneurship from the ivory tower: Do incentive systems matter? *Journal of Technology Transfer, 29*(3–4), 353–364.

Markman, G. D. S., Siegel, D. S., & Wright, M. (2008). Research and technology commercialization. *Journal of Management Studies, 45*(8), 1401–1423.

Mitchell, L. (2010). Testimony on "Improving Technology Commercialization to Future Economic Growth". Hearing U.S. House of Representatives Committee on Science and Technology, Subcommittee on Research and Science Education. June 10, 2010.

Mowery, D. C., Nelson, R. R., Sampat, B. N., & Arvids, A. Z. (2001). The growth of patenting and licensing by the U.S. Universities: An assessment of the effects of the Bayh-Dole Act of 1980. *Research Policy, 30*(1), 99–119.

Mowery, D. C., & Sampat, B. N. (2001b). University patents and patent policy debates in the USA, 1925–1980. *Industrial and Corporate Change, 10*(3), 781–814.

Mowery, D. C., Nelson, R. R., Sampat, B. N., & Arvids, A. Z. (2004). *Ivory tower and industrial innovation: University-industry technology transfer before and after the Bayh-Dole Act in the United States.* Stanford: Stanford Business Books.

Mowery, D. C., & Sampat, B. N. (2005). The Bayh-Dole Act of the 1980 and university-industry transfer: A model for other OECD governments? *Journal of Technology Transfer, 30*(1–2), 115–127.

National Institute of Health (NIH). (2012). NIH Budget. NIH Website. http://www.nih.gov/about/budget.htm. Accessed 1 Mar 2012.

National Research Council (NRC). (2003). *The impact of academic research on industrial performance.* Washington, DC: National Academy.

National Research Council (NRC). (2003b). *Innovation in information technology.* Washington, DC: National Academy Press.

National Research Council (NRC). (2010). Managing university intellectual property in the public interest. In A. M. Stephen & A. M. Mazza (Eds.), *Board on science, technology, and economic policy; committee on science, technology, and law; policy and global affairs.* Washington, DC: National Academy Press.

National Science Board. (2010). *Science and Engineering indicators: 2010.* Arlington: National Science Foundation.

National Science Foundation (NSF). (2000). The Nifty50. NSFwebsite. http://www.nsf.gov/od/lpa/nsf50/nsfoutreach/htm/n50_z2/pages_z3/text_list.htm. Accessed 1 Mar 2012.

Powell, W. W., & Owen-Smith, J. (1998). Universities and the market for intellectual property in the life sciences. *Journal of Policy Analysis and Management, 17*(2), 253–277.

Rhoten, D., & Powell, W. W. (2007). The frontiers of intellectual property. Expanded protection versus new models of open science. *Annual Review of Law and Social Science, 3,* 245–273.

Ritchie de Larena, L. (2007). The price of progress: are universities adding to the cost? *Houston Law Review, 43*(5), 1374–1444.

Roessner, D., Bond, J., Okubo, S., & Planting, M. (2009). The economic impact of licensed commercialized inventions originating in university research, 1996–2007. Final Report to the Biotechnology Industry Organization. September 3, 2009.

Rothaermel, F. T., Agung, S. D., & Lin J. (2007). University entrepreneurship: A taxonomy of the literature. *Industrial and Corporate Change, 16*(4), 691–791.

Rosenberg, N. (2003). America's Entrepreneurial Universities. In D. M. Hart (Ed.), *The emergence of entrepreneurship policy*. Cambridge: Cambridge University Press.

Sampat, B. N. (2003). Recent changes in patent policy and the 'privatization' of knowledge: Causes, consequences, and implications for developing countries. In D. Sarewitz (Ed.), *Knowledge flows & knowledge collectives: Understanding the role of science & technology in development*, (pp. 39–81). Washington, DC: Cent. Sci. Policy Outcomes. http://www.cspo.org/products/rocky/Rock-Vol1-2.PDF.

Schacht, W. H. (2009). The Bayh-Dole Act: Selected issues in patent policy and the commercialization of technology. Congressional Research Service Report for Congress. December, 16, 2009.

Schacht, W. H. (2011). The Bayh-Dole Act: Selected issues in patent policy and the commercialization of technology. Congressional Research Service Report for Congress. June, 9, 2011.

Severson, J. A. (2000). Oversight hearing on "gene patents and other genomic inventions". House Committee on the Judiciary. Subcommittee on Courts and Intellectual Property. U.S. House of Representatives. July, 13, 2000.

Shane, S. (2004). Encouraging university entrepreneurship? The effect of the Bayh-Dole Act on university patenting in the United States. *Journal of Business Venturing, 19*(1), 127–151.

Siegel, D. S., Waldman, D. A., Atwater, L. E., & Link, A. N. (2004). Toward a model of the effective transfer of scientific knowledge from academicians to practitioners: Qualitative evidence from the commercialization of university technologies. *Journal of Engineering and Technology Management, 21*(1–2), 115–142.

Smith, T., Samors, R., Heinig, S., & Hardy, R. (2010). Commercialization of university research. Memorandum to Office of Science and Technology Policy and National Economic Council on Federal Register Request for Information for Commercialization of University Research. May 10, 2010. Association of American Universities, Association of Public and Land-Grant Universities, American Council on Education, Association of American Medical Colleges, and Council on Government Relations. http://www.aplu.org/NetCommunity/Document. Doc?id=2515. Accessed 1 Mar 2012.

Thursby, J. G., Jensen, R. A., & Thursby, M. C. (2001). Objectives, characteristics and outcomes of university licensing: A survey of major U.S. universities. *Journal of Technology Transfer, 26*, 59–72.

Thursby, J. G., & Thursby, M. C. 2002. Who is selling the ivory tower? Sources of growth in university licensing. *Management Science, 48* (1), 90–104.

Thursby, J. G. & Thursby, M. C. (2006). University Licensing. Where is the new science in corporate R&D? *Science, 314*(5805), 1547–1548.

Thursby, M. C., Thursby, J. G., & Gupta-Mukherjee, S. (2007). Are there real effects of licensing on academic research? A life cycle view. *Journal of Economic Behavior & Organization, 63*(4), 577–598.

Valdivia, W. (2013). University start-ups: Critical for improving technology transfer. Brookings Center for Technology Innovation. http://www.brookings.edu/research/papers/2013/11/university-start-ups-technology-transfer-valdivia. Accessed 1 Sept 2014.

Van Looy, B., Ranga, M., Callaert, J., Debackere, K., & Zimmermann, E. (2004). Combining entrepreneurial and scientific performance in academia: Towards a compounded and reciprocal Matthew-effect? *Research Policy, 33*(3), 425–441.

Washburn, J. (2005). *University Inc: The corporate corruption of higher education*. New York: Basic Books.

Wisconsin Alumni Research Foundation. (2010). Our History. Wisconsin Alumni Research Foundation Website. http://www.warf.org/about-us/background/history/history.cmsx. Accessed 1 Sept 2014.

Youtie, J., & Shapira, P. (2008). Building an innovation hub: A case study of the transformation of university roles in regional technological and economic development. *Research Policy, 37*(8), 1188–1204.

Chapter 4
Strategic Alliances/Knowledge-Intensive Partnerships

Timothy Williams and Nicholas S. Vonortas

4.1 Introduction

The confluence of important developments in the international economic environment during the past two to three decades has turned inter-firm cooperation into an important mechanism of business interaction and market and technology access (Malerba and Vonortas 2009; Caloghirou et al. 2004; Jankowski et al. 2001; Vonortas 1997). Particularly in high- and medium-tech industries, the private sector has increasingly used various kinds of cooperative agreements such as joint ventures, joint R&D, technology exchange agreements, co-production, direct minority investments, and sourcing relationships to advance core strategic objectives. Called alliances (partnerships) in this chapter, such agreements imply deeper and steadier relationships than arm's-length market exchanges but fall short of complete mergers. They involve mutual dependence and shared decision-making between two or more independent parties. When research and development is a focus of the partnership, universities and other research institutes may also participate.

The proliferation of inter-firm alliances has raised expectations of accelerated long-term growth opportunities for developing countries through faster access to markets and technologies and greater learning possibilities. Available evidence, however, shows that although developing country firms have increased their participation significantly, recorded partnerships are still overwhelmingly concentrated in developed economies. It also shows that a rather small group of newly industrializing

Adapted from a chapter of the *Innovation Policy Handbook* report composed for the World Bank (2012). Original unpublished and available upon request.

T. Williams (✉)
Center for International Science and Technology Policy, The George Washington University, Washington, D.C., USA
e-mail: timnwilliams@gmail.com

N. S. Vonortas
Center for International Science and Technology Policy & Department of Economics, The George Washington University, Washington, D.C., USA
e-mail: vonortas@gwu.edu

© The Editor(s) 2015   47
N. S. Vonortas et al. (eds.), *Innovation Policy,* SpringerBriefs in Entrepreneurship and Innovation, DOI 10.1007/978-1-4939-2233-8_4

countries and economies in transition which have significant capabilities and large domestic markets have benefited disproportionately more than others.

Rather than equity-based, the vast majority of partnerships during the past 20 years have been contractual agreements, catering to the pressing need for strategic flexibility in high-tech sectors. Strong arguments can be made that non-equity agreements can favor of developing country firms as they require less commitment and get closer to informal kinds of cooperation. Numerous cases of transnational companies operating in developing countries and emerging economies have shown how cross-border partnering and networking can significantly raise those countries' technological prowess and business competitiveness.

Analysts may have paid too much attention to formal forms of partnering—like those mentioned above, involving explicit contracting among parties—and much less attention to various forms of informal partnering among organizations and individuals. Anecdotal evidence indicates that informal partnering probably accounts for a very large share of partnering activity in industry, involving extensively small and medium-sized enterprises (SMEs) in proximate geographical areas.

Formal and informal partnering should be seen as a continuum, where formal enterprise cooperation, clustering and networking are perceived as alternative, and often complementary, modes of operation. Formal partnership requirements—including strategy formulation and significant partner contribution in tangible and/or intangible resources—may be placing the bar too high for the majority of (mainly small) firms in most developing countries. That, however, leaves a whole lot of other cooperative interactions for these economic agents to pursue. It now seems quite probable that more informal partnering through networks and clusters is a way for many firms in developing countries to increase their sophistication and become stronger and more competitive, thus gradually preparing for more formal partnerships.

For firms that do graduate to formal partnerships, this Chapter expounds a roadmap to harnessing their potential for promoting technological prowess and economic competitiveness. Key lessons for success include a clear understanding of the firm's objectives in the partnership, the negotiation of a suitable agreement with sound dispute resolution and exit clauses, the treatment of the agreement as a "living" document, and the awareness of the importance of knowledge and relative capability distribution among partners. For these firms, policy decision-makers and international organizations have important roles to play in terms of spreading the message of partnership opportunities, on one hand, and in terms of creating a supportive infrastructure, on the other.

4.2   Common Types of Alliances

Three types of alliance are particularly common:

- **Equity shareholding**: Arrangement in which a company becomes a minority shareholder in its partner through an equity investment. This action is often reciprocated by the alliance partner.

- *Example*: In 1999, Renault and Nissan entered a strategic alliance through a cross-shareholding agreement, whereby each company purchased a minority equity stake in the other. Renault currently holds a 43.4% stake in Nissan while Nissan holds 15% of Renault shares. This arrangement ensures that each company will act in the financial and strategic interests of the other while maintaining its own identity and culture. Activities include joint production of engines, batteries, and other key components.

- **Joint Venture:** Arrangement in which partners agree to contribute resources and equity to develop a new business entity with a specific purpose in mind.

  - *Example*: In order to save money on procurement operations, in 2011 Deutsche Telekom (DT) and France Telecom (FT) created a new 50/50 joint venture firm known as BUYIN. The new company, which is based in Brussels, manages the procurement of terminal devices, mobile communications networks, and fixed network equipment for the two telecom giants. The alliance is expected to save the companies about € 1.3 billion over the first three years of operation. Furthermore, DT and FT have expressed interest in expanding the joint venture to other areas such as IT infrastructure in the future.

- **Contractual (non-equity):** Arrangement that lacks shared ownership or dedicated administrative structures. Cooperation is undertaken through non-equity based means such as licensing deals, technology exchange agreements, sourcing relationships, co-marketing, etc.

  - *Example*: Malaysia's AirAsia and Australia-based Jetstar teamed up in 2010 with a plan to reduce the two budget airlines' operating costs. Through a non-equity alliance, the airlines agreed to explore opportunities to jointly procure aircraft, cooperate in passenger handling in Australia and Asia, pool aircraft components and spare parts, and jointly acquire engineering and maintenance supplies and services. The airlines expect the alliance to reduce costs, pool expertise and result in cheaper fares.

Indus Towers Joint Venture

An example of a successful joint venture in the telecom industry is the Indian tower management company Indus Towers. Indus Towers was established in November 2007 through a joint venture between BhartiAirtel, Vodafone Essar, and Idea Cellular, with the goal of reducing passive infrastructure costs for each company. Over the past decade, the Indian telecom industry has been undergoing extraordinary growth, with some experts forecasting an 80% penetration rate by as early as 2017. Early competition in this industry was intense and marginal revenues were very low compared to other countries, which led to challenges with capital investment in new tower infrastructure. At the beginning of 2007, only 25% of wireless towers in India were shared between telecom operators. This system was inefficient for operators because firms were building towers in overlapping areas that could easily be serviced by a single tower.

BhatiAirtel and Vodafone Esser, the two largest private telecom-services providers in India, realized they could cooperate on tower development while remaining competitive in their core businesses of providing telecom services. Together, they decided to jointly establish an independent firm to construct and manage towers throughout the two firms' common operating regions. Idea Cellular, the third largest telecom operator in India, was also offered a smaller share in the new firm and eagerly accepted based on the expansion prospects it could provide.

Negotiating and implementing the terms of the joint venture included several challenges that needed to be resolved by the parties involved. Determining how to value the assets that each company contributed was an early area of friction, which was resolved through the establishment of a point system where towers were rated based on attributes such as location and size. The companies then contributed capital for new towers such that the point values were equal among each partner. Other early issues included network downtime, the lack of a standardized data sharing platform, and conflicts between strategic company objectives. In the face of these challenges, Indus Towers was able to find solutions in large part due to equal representation on the management board and a shared understanding of the challenge that needed to be solved.

Over the next 4 years, Indus Towers had grown into an efficient vehicle to operate towers throughout the country and had successfully evolved into an independent tower company. At the beginning of this decade Indus Towers was the largest telecom tower company in the world with a portfolio of over 110,000 towers and plans to add 5000 more each year until 2015.

Source: Gulati et al. 2010

4.3 Context of Strategic Alliances

4.3.1 Definitions

Alliances refer to agreements whereby two or more partners share the commitment to reach a common goal by pooling their resources together and by coordinating their activities. Partnerships denote some degree of strategic and operational coordination and may involve equity investment. They can occur vertically across the value chain, from the provision of raw materials and other factors of production, through research, design, production and assembly of parts, components and systems, to product/service distribution and servicing. Or, they can occur horizontally, involving competitors at the same level of the value chain. Partners may be based in one or more countries.

A narrower set of partnerships can be characterized as innovation-based, focusing primarily on the generation, exchange, adaptation and exploitation of technical advances. Called strategic technology alliances (STAs) herein, these arrangements are of primary concern to both developed and developing countries as a result of expected direct contribution to national capacity building.

The most basic distinction in partnerships is between formal and informal agreements. Relatively little is known about the latter apart from anecdotal evidence that (a) many firms routinely partner informally on short-term business endeavors, and (b) informal partnerships may account for the vast majority of collaboration. Informal partnerships are unfortunately almost impossible to track down systematically. They fall more in the realm of clusters and networks to which we will return in the last section.

4.3.2 International Context

Since the early 1980s, when the first data were put together to map a sudden burst of inter-firm cooperation, it has been established beyond doubt that alliances have become an important mechanism of business interaction and market and technology access around the world. A proliferating literature in economics, business and policy has tried to identify and interpret the important features of cooperation among firms, universities, and other public and private organizations.[1]

A set of developments in the international economic environment has underlined the explosion of business partnerships since the late 1970s. Four changes, in particular, seem to be key:

- *Globalization.* Transnational companies have pushed into new product and geographical markets relentlessly.
- *Technological change.* The pace of technological advance has accelerated significantly, partly as a result of increasing competition through globalization. In addition to being an outcome of competitive pressures, however, technology is an enabler of globalization. Technological capabilities have diffused around the world more widely than ever before.
- *Notion of "core competency".* Increasing international competition and faster pace of technological advance have robbed firms of their ability to be self-sufficient in everything they want to do. The current management mantra is to do internally what a company does best and outsource the rest through partnerships.

---

[1] For literature reviews see, for example, Caloghirou et al. (2003, 2004), Gomes-Casseres (1996), Gulati (1998), Hagedoorn et al. (2000), Hemphill and Vonortas (2003), Vonortas and Zirulia (2011).

- *Economic liberalization and privatization.* This process has led to unprecedented international flows of capital in the form of both foreign direct investment and portfolio investment. Developing countries have managed to increase their share of the intake (but the distribution among them remains highly skewed).

Such developments have changed the nature of international business interactions that has supported the development of a score of developing countries since the mid-twentieth century. Traditional mechanisms of technology transfer including licensing, the acquisition of capital goods, and the transfer of complete technology packages through foreign investment are being supplemented by many semiformal and formal new mechanisms for gaining access to technologies and markets. These new mechanisms entail the formation of dense webs of inter-organizational networks that provide the private sector with the necessary flexibility to achieve multiple objectives in the face of intense international competition. The result has been an increasing interdependence on a global scale that few firms interested in long-term survival and growth can escape.

The available literature on formal business partnerships and networking has tended to focus primarily on developed countries: their firms have dominated global partnering records, at least as currently accounted for. OECD member countries have accounted for no less than four-fifths of the activity over the years. More recently the rapidly developing economies of China, India, and Brazil have registered significant international cooperative activity, especially large multinational corporations based in these countries. The same firms also dominate international trade and investment.[2]

The vast majority of the recorded alliances are classified as contractual agreements. Contractual agreements do not involve equity investment across partners or in the collaborative activity (such as in a joint venture). Sectors registering large numbers of partnerships around the globe include pharmaceuticals, chemicals, electronic equipment, computers, telecommunications, and financial and business services. Service sectors took an increasing share of the total in more recent years. The motives of firms to partner differ among sectors. Cost-economizing—e.g., share costs and risks of a technological development—appears to be more significant in capital and R&D intensive sectors such as telecommunication hardware. Strategic considerations become important when firms use partnerships to enter new product areas, especially ones with high technological and market risk. In information and communication industries a major driving force towards international partnerships seems to be the effort to develop new global product and system standards. In pharmaceuticals, cost economizing and speed to market seem to be very important. In the automotive sector, securing resources to develop state-of-the-art technologies for environmental friendly vehicles, achieving economies of

---

[2] For references to partnering in developing and transition countries see Deloitte (2004), Freeman and Hagedoorn (1994), Ivarsson and Alvstam (2005), Lee and Beamish (1995), Rondinelli and Black (2000), Si and Bruton (1999), and Vonortas (1998). A series of publications by UNCTAD review the literature on partnering and networking for national capacity building (UNCTAD 1999a, 1999b, 2000a, 2000b).

scale in production, and accessing markets appear to be major drivers. Finally, in the airline industry cost savings through investment in common systems of reservations, ticketing, and client services appear to be the main driving force for international partnering activity.

A major development has been the contrasting evolution of equity-based STAs (e.g., traditional joint ventures) and non-equity STAs in the past two decades. From almost 100% in the mid-1960s, the share of equity-based STAs in the total fell to about 70% in the 1970s, 40% in the 1980s, less than 20% in the 1990s, and less than 10% more recently. The gap has been filled by non-equity, contractual forms of STAs such as research consortia and joint development agreements that have provided the main mechanism of inter-firm collaboration in more recent years. For instance, all countries with significant public R&D programs fund research consortia these days, with the most prominent example being the Framework Programmes for Research and Technological Development of the European Union.

High-tech manufacturing sectors—information technology, pharmaceuticals, aerospace, defense—have gradually developed a dominant position in STAs since the early 1980s. Medium-tech sectors—instrumentation and medical equipment, automotive, consumer electronics, chemicals—have followed. High-tech sectors have strongly preferred contractual STAs, relative to medium- and low-tech sectors.

4.4 A Practical Guide

Alliances can significantly expand opportunities for companies interested in accessing markets and technologies and for governments interested in indigenous capacity building and economic growth. However, benefits do not flow automatically; nor do partners necessarily gain equally. There is a lot of learning associated with setting up and managing successful partnerships and room for policy decision making to facilitate them. This section distills lessons from past experience to draw a practical generic guide to negotiating and managing successful partnerships. It focuses mostly on STAs.

4.4.1 Partnership Opportunities and Dangers

Consideration of a business partnership must always start with a careful recount of the *strategic challenges* confronting the firm in question. Management must consider:

- Where does the firm want to go in the future? What are its strategic objectives?
- What are the necessary projected steps—organizational, technological, finance, marketing, and so forth—to achieve the strategic objectives?
- To what extent do the required resources and capabilities exist internally?

The more *tactical challenges* for management considering a specific task include:

- What is the exact activity the firm is currently interested in and why can it not be either carried out in-house or bought from an external source?
- How is a partnership expected to assist in accessing the requisite resources and capabilities that the firm does not already possess?
- What kind of partners is the firm interested in? How is it going to identify them?
- How to successfully negotiate the partnership? What are the specific assets that the firm will bring to the negotiating table? How much control can it afford to give away?
- How to manage the partnership and learn from it?
- How to set clear objectives for the partnership?
- How to evaluate partnership performance?
- When and how to dissolve the partnership?

From the point of view of the firm, potential *benefits from partnering* include:

- Access to markets; create new product markets;
- Share costs of large investments;
- Share risk, reduce uncertainty;
- Access complementary resources and skills of partners, such as complementary technologies, people, finance; exploit research and technological synergies;
- Accelerate return on investments through a more rapid diffusion of assets;
- Rationalize the deployment of resources to enhance economies of scale and scope;
- Increase strategic flexibility through the creation of new investment options;
- Unbundle the firm's portfolio of intangible assets, and selectively transfer components of this portfolio;
- Co-opt competition;
- Attain legal and political advantages in host countries.

More broadly, alliances have such virtues as flexibility, speed, and economy. They can be put together in little time and be folded up just as quickly. They can involve little paperwork. An analogy of partnerships *vis-a-vis* market internalization through mergers and acquisitions would be "love affairs" rather than "marriages".

Alliances also entail costs. Regardless of strategic goals, inter-firm collaboration always implies a *trade-off* between greater access (markets, finance, resources, and capabilities) and lesser control of strategic decision making, day-to-day management, technological and other kinds of proprietary knowledge. Partial loss of control over strategic decisions, over technology use, and over market position can invite opportunistic behavior by one or more partners resulting in the involuntary loss of important assets, particularly intangible assets such as technological and other types of knowledge. Other potential *drawbacks from partnering* include:

- Increased transaction costs due to

  - increased management needs,
  - diversion of management attention
  - employee coaching into the agreement

- decisions and responsibilities that are subject to negotiation.

• Lack of compatibility of the collaborative activity with core firm interests; e.g., locking the firm into a product/service standard that may not be in its best interest.

It should be stressed that partners often join a partnership for different reasons. Reasons for participation can shift over time, implying shifts regarding the perceived benefits and costs of collaboration. The motivation to enter into a joint relationship must, then, be not only strong but regularly reexamined during the lifetime of the partnership.

Petrobrás Subsea Boosting Technology Development

Over the last several decades, Brazil's Petrobrás has evolved successfully into a global leader in deep sea drilling techniques by using strategic alliances to help it absorb external knowledge and generate unique solutions. Particularly, the alliance strategies that it employed during the 1980s and 1990s played a crucial role in its development of subsea boosting technologies.

Subsea boosting refers to technologies that increase the flow rate of wells in deep sea oil fields. This has been an important area of concern for Brazil since most of its recent large oil discoveries have been found under these conditions. Before Petrobrás utilized subsea production, it was limited to using a Floating Production System (FPS) which was subject to problems including limited depth capabilities and setbacks due to poor weather.

Petrobrás' development of Subsea Multiphase-flow Pumping Systems (SBMS) showcases how it navigated these challenges to join the select club of firms that operate subsea production systems. It began with little to no knowledge of the technology, but was able to join an industry project to research SBMS technology, led by Scottish pump manufacturer Weir Pumps. Petrobrás' role in the project was limited due to its lack of experience, but it was able to use this experience to monitor the progress in SBMS technology and understand new developments that occurred. The project ultimately failed, but Petrobrás succeeded in gaining a much deeper knowledge of the hurdles facing the technology and which competing avenues held promise. This knowledge helped Petrobrás take the next step and establish a technological cooperation agreement with German pump manufacturer Borneman, with the goal to develop a prototype system that was suited for utilization in Brazil's offshore fields. It took a much more active role in this project and contributed extensively to a testing campaign that identified and ultimately solved the bottlenecks in the system. By 1997, Petrobrás was ready to put the innovation into production. At this time, Petrobrás ended its relationship with Borneman and entered a new joint industry project in which Westinghouse, Leistritz, and a host of other suppliers would take part in delivering the system to Petrobrás. The decision to shift away from Borneman was purely an economic choice. Petrobrás had already acquired the technological know-how it needed to implement the system and became more concerned with system costs than technology development.

The experience of Petrobrás in its development of SBMS systems highlights how it used different modes of partnering at different stages of development in order to attain the maximum benefit at each stage. In the first stage, it was mainly concerned with learning about opportunities, and the joint industry project served as an entry point to monitor progress in the sector while minimizing costs to the firm. From here, Petrobrás was able to develop its own technology through a technology cooperation agreement and ultimately mastered this technology. Finally, it commercialized this technology through the use of industry collaboration in order to reduce its costs. Although the Petrobrás experience is special due to the great amounts of capital available to the company, it illustrates how partnering is a fluid endeavor with requirements that change and evolve as a firm progresses towards its objective.

Source: Furtado and Gomes de Freitas 2000

Tata-Fiat Joint Venture

The challenges of developing a successful joint venture are exemplified by the partnership between the Italian automaker Fiat and its Indian partner, Tata Motors. In 2007, the companies created a joint venture firm to produce engines, transmissions, and complete automobiles at plants in India. With a strong relationship previous to the agreement, the JV firm seemed like a natural progression for two companies with similar values and objectives. Fiat already had a presence in India for several decades, and established a wholly-owned subsidiary, Fiat India, in 1995. However, the Indian subsidiary struggled in the following decade, leading company executives to believe the company could not "go it alone" in the Indian market. They felt that Fiat needed a committed partner to identify appropriate products and prices for the Indian market, build an effective distribution network, and commit to a long-term arrangement. Tata Motors, on the other hand, was in a position to benefit from Fiat's technical expertise and global business network.

In 2005, the two companies began a dialogue on how they could mutually benefit from cooperation. Through high-level discussions, Fiat and Tata executives soon realized that the companies had much to gain from one another. The meetings soon led to a Memorandum of Understanding, which solidified their intent to "analyze the feasibility of cooperation, across markets, in the area of passenger cars that would encompass development, manufacturing, sourcing and distribution of products, aggregates and components." A year later, the two companies signed an agreement for a dealer sharing network in India, with Tata Motors managing the marketing and distribution of two Fiat models, the Palio and Palio Adventure. Soon thereafter, the head of Tata Motors, Ratan Tata, was appointed to the board of directors of Fiat, signaling a new era of cooperation between the firms. This increasing level of integration set the stage for the 50-50 joint venture, which was agreed upon after a long negotiation process involving aspects such as asset values and exit clauses. The agreement seemed at first to be a golden opportunity for both firms.

Four years later, the alliance between Fiat and Tata was still in operation, with a good number of vehicles produced by the joint venture since its inception. However, as of 2011 the partnership had yet to break even and was increasingly on shaky ground. Fiat's product line had struggled to gain ground in India, with many analysts pointing a lack of Fiat model variety, and a poor perception of Fiat in India generally as the source of strains. Still, the challenges associated with the partnership may run deeper than product lineup and marketing failures. Many cultural differences exist not only on a corporate level, but on a national level as well. The future of the Fiat-Tata alliance was still uncertain, but one thing had become clear: executives from both firms should work together to improve Fiat's image and appeal in the Indian marketplace if the venture was to succeed in the long-run.

Sources: Mitchell et al. 2008; Chaudhari 2011

4.5 Partner Choice

The existence of complementary needs, assets, and capabilities among partners is generally considered a prerequisite for maximizing collaboration benefits and minimizing costs. Complementarities may be reflected in:

- Expertise in different, but commercially linked, technologies;
- Strength in different, but commercially linked, markets;
- Specialization in separate parts of the value chain.

The trade-off of linking complementary organizations may be higher transaction costs for running the partnership. The chance for disagreements, for instance, between partners on market strategy, technology designs, and decision-making processes rises. Holding all else constant, like-minded partners with similar management perspectives, goals and will result in fewer conflicts and lower costs of managing collaboration.

Common Alliance Problem: Choosing the Wrong Partner

The risks involved in strategic alliances increase substantially when the alliance is codified in a written contract, and especially when there is uncertainty about the future or a partner's reliability. For example, when Dow Chemicals signed a $ 17.4 billion Joint Venture Formation Agreement with Kuwait's state-run Petrochemical Industries Company (PIC) in 2008, everything seemed to be on track for the creation of a new leading global plastics manufacturing company known as K-Dow. Shortly after the 50-50 joint venture deal was inked, however, PIC's parent company, Kuwait Petroleum Corporation, reneged on the agreement with concerns over the ensuing global recession.

The breakup of the joint venture agreement had severe consequences for Dow, which had expected $ 7.5 billion in revenue from the sale of several chemical plants to PIC. Prior to the debacle, Dow had agreed to acquire a rival firm, Rohm and Haas, with the funds it had planned on receiving from the joint venture deal. Not only did the failure of the venture lead to a drawn out legal battle between Dow and PIC, but Dow was also facing a lawsuit from Rohm and Haas for failing to honor the acquisition deal.

Sources: Sieb 2008; Westervelt 2009

4.6 Partnership Negotiation[3]

Negotiation is one of the most important aspects of partnerships. Depending on the objectives, experience, and complexity of the deal, partnership negotiation can be a difficult process. The length of negotiation is reported to vary from a few weeks up to 2 years. Several issues are extremely important and tend to dominate the negotiation phase:

- Control of the partnership, including its equity structure and veto power over various aspects in managing the partnership (appointment of key personnel, dividend policy, technology use, export markets, quality standards, supply sources, etc.);
- Conditions surrounding technology transfer. This is the most frequently mentioned item in partnership contracts following control;
- Dispute resolution;
- Terms of partnership termination.

Common negotiation problems include:

- Valuation of the assets brought by each partner to the partnership;
- Transparency;
- Conflict resolution procedures—explicit rules and/or trust relationships;

---

[3] The section draws considerably on Miller et al. (1995).

- Allocation of management responsibility and degree of management independence;
- Changes in ownership shares as partnership matures;
- Exit policy;
- Dividend policy;
- Measurement of performance.

Managing Alliances: Eli Lilly's Corporate Strategy

In 1999, Eli Lilly established the pharmaceutical industry's first "Office of Alliance Management" which was established specifically to implement and guide alliances once agreements are made. Eli Lilly's management recognized that most unsuccessful alliances fail due to implementation issues, personality conflicts and other non-technical factors. The Office of Alliance Management addresses these issues and works closely with partners to ensure strategic, operational, and cultural alignment to optimize resources and meet alliance goals. This office is part of a larger framework of Eli Lilly's alliance building strategy, which also includes offices geared towards identifying opportunities and negotiating agreements with partners.

Source: Stach 2006

Fairly common *relationship problems* include:

- *International strategy-related problems.* A particular type of conflict in cross-border alliances may occur when a multinational corporation (MNC) with a global strategy forms a partnership with a local partner pursuing more narrowly defined goals. Global strategies frequently require the MNC to incur costs in one country in return for profits in another. Local partners may thus be placed at a disadvantage. Given that relationships can shift over time, this may become a problem during the course of the partnership. Such problems can include the following:

  - Export rights. Exporting sometimes represents a fundamental difference between industrial and developing country partners. A MNC may not want the partnership to freely export products to markets already be served from other manufacturing points in its system. The developing country partner will be of a different opinion as it will typically view exports as a natural avenue of expansion.
  - Tax issues. The optimization process undertaken by the MNC will cover its worldwide burden. If the partnership exports products through the TNC system, transfer-pricing strategy will not necessarily be in the interest of the local partner.
  - Dividend, investment policies. The global investment programs of the MNC may affect its preference of dividends over reinvestment in the partnership. Again, the local partner may have diverging views.

- Partner size. Large size differences may introduce difficulties during rapid expansion periods of the partnership due to their different resource base. Size differences may also have operational implications that can cause problems (e.g., the larger firm not taking the partnership seriously enough).

- *Ownership and control problems.* Long-term, strategic partnerships may need operational management with considerable independence from either partner. Problems may arise from changes during the lifetime of the partnership. A possible change involves the management in one of the partners that may affect this firm's attitude towards the specific partnership. In addition, one needs to consider possible disagreements over time regarding changes in product lines, raw material sourcing, technology transfer and utilization, and so forth.
- *Cultural problems.* These involve both the social cultural backgrounds of companies based in different countries and the corporate culture that characterizes each company. Both types of cultures condition how people view their environment and how they interpret issues. Complaints concerning arrogance, business practice, corruption, and so forth have not been unknown to partnerships.
- *Problems related to dynamic changes in the relationship.* The changing environment within which the partnership operates alters partner relationships and can cause stress.

  - Experience in a partnership results in learning. Learning can modify how one views the contributions of the partner. Learning should happen from all sides and involves better market understanding and improved capabilities. Learning boosts self-confidence and raises expectations for partner contribution. The result sometimes is dissatisfaction. Moreover, dissatisfaction is frequently the result of differential rates of learning that make a firm feel falling behind its partners.
  - Unforeseen changes in circumstances making parts of the agreement obsolete. Introducing the necessary modifications may be difficult, even in cases where all sides agree.

Common Alliance Problem: Differential Rates of Learning

Looking to expand into the Japanese marketplace in the 1970s, General Foods Corporation entered a partnership with Japanese food giant Ajinomoto. Ajinomoto offered its marketing expertise and knowledge of local business practices in Japan, and General Foods agreed to disclose its advanced processing technology for products such as freeze-dried coffee. After several years of successfully partnering together, Ajinomoto's management began to feel that the alliance was unnecessary because Ajinomoto had internalized the advanced processing technology and was no longer learning from its American partner. General Foods, however, was not as successful learning about the Japanese marketplace and still needed Ajinomoto's expertise. When the collaboration deteriorated and eventually disbanded, General Foods was left disappointed.

Source: Barlett et al. 2008

4.7 Conclusion

The proliferation of partnerships during the past three decades has raised expectations of accelerated growth through faster access to markets and technologies and greater learning possibilities. There is evidence that inter-firm partnerships can be an extremely useful tool to assist developing country firms in their efforts to catch up. Partnerships can accordingly assist countries speed up the process of establishing competitive indigenous industries. Partnerships can also play a major role in mobilizing the necessary resources and technological expertise to upgrade lagging infrastructure.

Formal partnerships require strategy formulation and partner contribution, whether in financial resources, intangible assets, market familiarity, market access, etc. Frequently, the required level of strategy sophistication and resource commitment is considerable. It is, thus, possible that these requirements raise the bar too high for the mass of (mainly small and unsophisticated) firms in the majority of developing countries. Still, this leaves many other interactions for these agents to pursue. It seems quite probable that informal partnering through networks and clusters is a way for many relatively disadvantaged developing country firms to become stronger, more competitive, and to meet the minimum capability prerequisites in order to graduate to formal partnerships. Governments may be wise to try addressing most developing country small firm problems related to size and competitive position through networks (often more vertical, supplier-buyer relationships) and clusters (regional, more horizontal, agglomerations).

For firms that do graduate to formal alliances, the following are key lessons for success:

- Clearly understand the strategic objectives of the firm.
- Clearly determine the firm's needs from the partnership.
- Negotiate a suitable agreement.
- Treat the partnership agreement as a "living" document.
- Understand that the comparative advantages of partners at the outset of the agreement may change over time.
- Be aware that technology transfer is one of the most sensitive and contentious issues. Create clear provisions for a framework of technology use in the partnership.
- Partnership agreements must contain sound provisions for dispute resolution and, in the event of irreconcilable differences, the exit mechanism to be employed in terminating the partnership.
- Monitor and review the partnership throughout its lifetime.

References

Barlett, et al. (2008). *Transnational management: Text, cases and readings in cross-border management* (5th ed.). New York: McGraw-Hill.

Caloghirou, Y., Ioannides, S., & Vonortas, N. S. (2003). Research joint ventures: A critical survey of theoretical and empirical literature. *Journal of Economic Surveys, 17*(4), 541–570.

Caloghirou, Y., Ioannides, S., & Vonortas, N. S. (Eds.). (2004). *European collaboration in research and development: Business strategies and public policies*. Northampton: Edward Elgar.

Casas, R., & Luna, M. (1997). Government, academia and the private sector in Mexico: Towards a new configuration. *Science and Public Policy, 24*(1), 7–14.

Casas, R., de Gortari, R., & Santos, M. J. (2000). The building of knowledge spaces in Mexico: A regional approach to networking. *Research Policy, 29*, 225–241.

Chaudhari, Y. (2011). Tata hints at rethink on joint venture with Fiat, DNA India. http://www.dna-india.com/money/report-tata-hints-at-rethink-on-joint-venture-with-fiat-1555071. Accessed 3 Nov 2014.

Deloitte Touch Tohmatsu. (2004). Partnerships for small enterprise development, report prepared for UNDP and UNIDO.

Freeman, C., & Hagedoorn, J. (1994). Catching up or falling behind: Patterns of international interfirm technology partnering. *World Development, 22*(5), 771–780.

Furtado, A. T., & Gomes de Freitas, A. (2000). The catch-up strategy of petrobrás through cooperative R&D. *The Journal of Technology Transfer, 25*(1):23–26 doi:10.1023/A:1007882903341.

Gomes-Casseres, B. (1996). *The alliance revolution: The new shape of business rivalry*. Cambridge: Harvard University Press.

Gulati, R. (1998). Alliances and networks. *Strategic Management Journal, 19*, 293–317.

Gulati, R., de Asis Martinez-Jerez, F., Narayanan, V. G., & Rachna, T. (2010). Indus Towers: Collaborating with Competitors on Infrastructure. Harvard Business School Case 110-057.

Hagedoorn, J. (2001). Inter-firm R & D partnerships: An overview of major trends and patterns since 1960. In J. Jankowski, A. N. Link, & N. S. Vonortas (Eds.), Strategic research partnerships. Workshop Proceedings, Arlington, VA: National Science Foundation.

Hagedoorn, J., Link, A. N., & Vonortas, N. S. (2000). Research partnerships. *Research Policy, 29*(4–5), 567–586.

Hemphil, T., & Vonortas, N. S. (2003). Strategic research partnerships: A managerial perspective. *Technology Analysis and Strategic Management, 15*(2), 255–271.

Humphrey, J., & Schmitz, H. (1995). Principles for promoting clusters and networks of SMEs, Paper #1, Small and Medium Enterprises Branch, United Nations Industrial Development Organization, Vienna: UNIDO.

Ivarsson, I., & Alvstam, C. G. (2005). Technology transfer from TNCs to local suppliers in developing countries: A study of AB volvo's truck and bus plants in Brazil, China, India, and Mexico. *World Development, 33*(8), 1325–1344.

Jankowski, J. E., Link, A. N., & Vonortas, N. S. (Eds.) (2001). *Strategic research partnerships*. Arlington: National Science Foundation.

Kang, N.-H., & Sakai, K. (2000). *International strategic alliances: Their role in industrial globalization, STI Working paper 2000/5*. Paris: OECD.

Lee, C., & Beamish, P. W. (1995). The characteristics and performance of Korean joint ventures in LDCs. *Journal of International Business Studies, 26*(3), 637–654.

Levitsky, J. (1996). Support systems for SMEs in developing countries: A review, Paper #2, Small and medium enterprises branch, United Nations Industrial Development Organization, Vienna: UNIDO.

Malerba, F., & Vonortas, N. S. (Eds.). (2009). *Innovation networks in industries*, editor, Edward Elgar.

Miller, R. R., Glen, J. D., Jaspersen, F. Z., & Karmokolias, Y. (1995). *International joint ventures in developing countries: Happy marriages? Discussion Paper #29, International Finance Corporation*. Washington, D.C.: The World Bank.

Mitchell, J., Hohl, B., Ariño, Martín, A., & Ozcan, P. (2008). Fiat's Strategic Alliance with Tata. IESE Business School, Case SM-1528-E.

Perez-Adelman, P. (2000). Learning, adjustment and economic development: Transforming firms, the State and associations in Chile. *World Development, 28*(1), 41–55.

Rondinelli, D. A. & Black S. Sloan (2000). Multinational strategic alliances and acquisitions in Central and Eastern Europe: Partnerships in privatization. *Academy of Management Executive, 14*(4), 85–98.

Si, S. X., & Bruton, G. D. (1999). Knowledge transfer in international joint ventures in transitional economies: The China experience. *Academy of Management Executive, 13*(1), 83–90.

Sieb, C. (2008). Kuwait decision to quit joint venture puts Dow Chemical's Expansion in Jeopardy. The Times, December 30, pp. 39

Stach, G. (2006). Business alliances at Eli Lilly: A successful innovation strategy. *Strategy & Leadership, 34*(5), 28–33.

United Nations Conference on Trade and Development. (1999a). *Working group on science and technology partnerships and networking for national capacity-building, Economic and Social Council, Commission on Science, Technology and Development, E/CN.16/1999/2.* Geneva: UNCTAD.

United Nations Conference on Trade and Development. (1999b). Report of the expert meeting on the impact of government policy and government/private action in stimulating inter-firm partnerships regarding technology, production and marketing with particular emphasis on North-South and South-South linkages in promoting technology transfers (know-how, management expertise) and trade for SME development, Trade and Development Board, Commission on Enterprise, Business Facilitation and Development, TD/B/COM.3/12, Geneva: UNCTAD.

United Nations Conference on Trade and Development. (2000a). Development Strategies and Support Services for SMEs: Proceedings of Four Intergovernmental Meetings, New York and Geneva: United Nations.

United Nations Conference on Trade and Development. (2000b). TNC-SME Linkages for Development: Issues—Experiences—Best Practices, New York and Geneva: United Nations.

United Nations Industrial Development Organization. (2001). *Development of clusters and networks of SMEs, report, private sector development branch.* Vienna: UNIDO.

Vonortas, N. S. (1998). Strategic alliances in information technology and developing country firms: Policy perspectives. *Science, Technology & Society, 3*(1), 181–205.

Vonortas, N. S. (2002). Building competitive firms: Technology policy initiatives in Latin America. *Technology in Society, 24,* 433–459.

Vonortas, N. S. (2007). *Cooperation in research and development.* Boston: Kluwer Academic Publishers.

Vonortas, N. S., & Safioleas, S. P. (1997). Strategic alliances in information technology and developing country firms: Recent Evidence. *World Development, 25*(5), 657–680.

Vonortas, N. S., & Zirulia, L. (2011). Business network literature review and building of conceptual models of networks and Knowledge-Intensive Entrepreneurship, Working Paper for the project "Advancing Knowledge-Intensive Entrepreneurship and Innovation for Economic Growth and Social Well-being in Europe", European Commission, DG Research and Innovation.

Westervelt, R. (2009). Dow launches arbitration proceedings against PIC. *Chemical Week, 171,* 7.

Chapter 5
Clusters/Science Parks/Knowledge Business Incubators

Benjamin B. Boroughs

5.1   Introduction

As information and communications technology (ICT) grew more advanced during the 1990s, some observers predicted that geographic location would cease to be a determining factor in economic development. In the old economy, factories had to be near raw materials like coal or iron ore. In the new economy, business would be global, with workers across the globe engaging with one another via mobile devices and the Internet. Instead, the last 20 years have shown that location still matters. While some services like call centers have been outsourced, they have been outsourced to particular places, like Bangalore in India, where many companies compete for business within a geographically restricted space. With this realization, economic development is now focused on creating local and regional agglomerations with a special focus, often aimed at the high-technology sector which is perceived to have high growth and export potential. This chapter focuses on these agglomerations, called clusters, and two policy options for encouraging high-tech growth, Science Parks and Knowledge Business Incubators. Despite the fact that many parks and incubators remain limited in scope, policy makers sometimes view such subsidized initiatives as the first seeds or stages of an economic continuum leading ultimately to the emergence of a vibrant high-tech cluster with many profitable private firms.

Adapted from a chapter of the *Innovation Policy Handbook* report composed for the World Bank (2012). Original unpublished and available upon request.

B. B. Boroughs (✉)
Center for International Science and Technology Policy, The George Washington University, Washington, D.C., USA
e-mail: benboroughs@gmail.com

© The Editor(s) 2015
N. S. Vonortas et al. (eds.), *Innovation Policy,* SpringerBriefs in Entrepreneurship and Innovation, DOI 10.1007/978-1-4939-2233-8_5

5.2 Clusters

In the second quarter of 2011, the Silicon Valley Region of the US State of California captured 39% of the roughly $7.5 Billion in US venture capital funding in that quarter. In a nation as vast as the United States, how did one relatively small geographic region, far from the financial and political centers of the US East Coast come to play such an important role in technology and innovation? The answer is that Silicon Valley is a phenomenally successful high-tech industrial cluster. Promoting cluster formation remains a common yet frequently elusive goal among technology and industrial policy makers across the world.

5.2.1 What is a Cluster and Why are they Desirable?

Just as moving people from a dispersed rural setting, to a dense urban one increases interaction and economic efficiency, so does concentrating businesses and specialists in one region increase their productivity and innovation. Michael Porter (1998) offers this succinct definition of clusters:

> Geographical concentrations of interconnected companies, specialized suppliers, service providers, firms in related industries, associated institutions (for example universities, standards agencies, and trade associations) in particular fields that compete but also cooperate.

More generally, clusters are agglomerations of people, firms, institutions, and other economic actors working in a similar field who interact in a relatively small region. While this Chapter focuses on high-tech clusters, such as in the fields of biotechnology and information and communication technology, low-tech clusters can also be extremely important economic drivers.

Indeed economic dynamism and innovation are precisely the qualities that attract policy makers to aid cluster-formation. High paying jobs, high economic growth, market dominant companies with export potential, and the prestige of being an international technological leader, are just some of the reasons high-tech clusters are so valued. A cluster can become a global center for the activity performed there, drawing investment from across a nation and the world. Examples of such dominant clusters range from financial services (Manhattan, City of London); shipping (Athens, Singapore); fashion (Milan, Paris); film and entertainment (Hollywood, Mumbai). High-tech clusters include electronics and software like Silicon Valley or biotechnology like Route 128 in Boston. Often high-tech clusters draw on the talent of top universities in the previous examples, Stanford and UC Berkeley and MIT and Harvard respectively.

Clusters are often described geographically, but it is not merely the proximity of related firms and institutions which makes them successful. It is the social interaction between economic actors which helps to drive innovation. A university may contain a brilliant scientist, a firm may retain a skillful lawyer or engineer, and a banker may possess access to great sums of capital, but if they never meet and discuss the ways that each may help the other a new innovative company is unlikely to be formed. In successful clusters, such collaboration and entrepreneurialism is profitably fostered.

Does an Innovative Cluster Need to be High Tech?

For the vast majority of developing countries it would be foolhardy to literally try creating "The Next Silicon Valley". It is not necessary to go after a leading edge high tech field such as software, biotechnology, or advanced materials to be innovative. Applying new technologies to older industries and encouraging an environment of collaboration, competition, entrepreneurship while extremely difficult, can boost the competitiveness of a region. One example is the Sinos Valley region of Brazil, which has grown from a regional center of shoe production into a major global exporter of shoes. Firms there have developed strong ties between firms, suppliers, and international retailers; this has dramatically increased the efficiency and scope of production (Nadvi 1995).

5.2.2   Why Do Industries Cluster?

When many businesses of the same type gather in one region, information sharing between firms, competition, and specialization spur development. A virtuous cycle develops where people seeking to be at the forefront of their field choose to live in the leading cluster and large talent pools in turn attracts more businesses. Workers then are even more likely to move to such an area because they are confident of finding employment and so on. Specialized financial institutions, tailored to a particular industry emerge, making business transactions easier. Increasing rates of return and positive externalities are key features of clusters. (Breschi and Malerba 2005).

Clustering also occurs because of the characteristics of four different kinds of knowledge relative to spatial proximity. These knowledge types are sometimes simplified as "Know-what", "Know-why", "Know-how", "Know-who". The first, "Know-what", refers to an up to date understanding of the state of the field. Both with regard to technology and changing business conditions; a firm grasp of formal and informal business and science news and facts. Know-what is needed to understand what direction companies should be moving in and is critical for strategic planning.

Analytical or scientific knowledge makes up "Know-why" which can be thought of as explanation of the works of nature. Both "Know-what" and "Know-why" are codifiable, that is, they refer to knowledge amenable to being written down, codified, and transmitted. Thanks to modern communication technology, codified

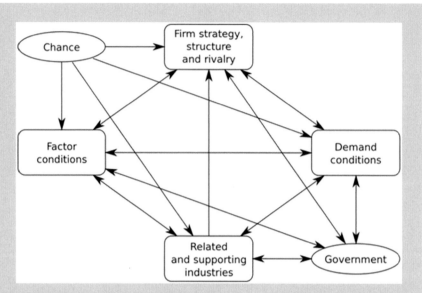

Fig. 5.1 Porter's diamond model (Porter 1998)

Porter (1998) popularized the Diamond Model as a way to analyze a region's strengths and weaknesses. Factor conditions refer to a region's inherent properties, such as skilled labor, access to capital, natural resources, and institutions. Demand conditions describe the structure of a region's home market. If the region's home market contains many sophisticated consumers of a technology, the region will be at an advantage because of the rapid market feedback they can receive. The web of supportive and related industries can also play a key role for the emergence of a cluster. Companies with active and engaged suppliers are more likely to innovate. Firm strategy, Structure, and Rivalry define how firms in a regional cluster will relate to one another. Collaborative, open relationships can speed the transfer of knowledge among market participants, but rivalry can also spur innovation through competition.The government can influence all aspects of market environment through its use of regulations, subsidies, taxes, education policy. Finally, chance can heavily influence the developmental trajectory and cannot be fully controlled by either firms or the government (Fig. 5.1).

knowledge can be transmitted around the world in a matter of seconds. Imagine a racing automobile; there is a great deal of information which can be transmitted about its qualities, specification, and care. This information can be found in blueprints, owner's manuals, cost invoices, and in detailed engineering test data. However, one would be hard pressed to take all this data and put together a championship Formula One racing team from even the most intelligent and athletic group of people unfamiliar with auto racing.

This is because a third kind of knowledge the "Know-how" is also critical. Tacit knowledge, also referred to as "learning through doing", is not easily transferred over long distances. Such knowledge, like the ability of a mechanic to instantly diagnose an unusual engine problem or a driver to know exactly how much to engage the clutch when approaching a hairpin turn cannot be appropriated through reading a book. Tacit knowledge is said to be "sticky" not moving fast or far from those who have it. Many industrial processes involve a great deal of tacit knowledge. Only by working side by side or closely collaborating can individuals fully master the ability to efficiently complete certain tasks.

Finally, "Know-who" refers to who knows how to do what, that is, information linking individuals and organizations to particular pieces of knowledge. Put differently, networking is the intimate knowledge of which individuals are truly important as innovators and institutional gatekeepers. Reputations can be difficult to judge from afar. Media sources may report on scientists who are the most interesting to readers or "colorful" while ignoring those in the field who are truly driving progress. Similarly in government or corporate bureaucracies, someone who holds a certain high rank or title may not actually be the key to an organization's management.

Location makes a significant difference for the application of all four types of knowledge. While tacit knowledge and networking are most obviously tied to geography, it turns out that much of analytical knowledge is as well. A study of research cited in patents, for instance, reveals that papers from nearby universities are more likely to be cited than papers from universities located farther away (Fagerberg et al. 2005).

5.2.3 Agglomeration Vs. Innovative Clustering

Cities have long contained districts which cater to a specific type of industry. Sometimes this occurred because of deliberate policy—grouping all butchers and abattoirs in one block to separate the process of animal slaughter from the rest of the city. Often though, and especially as modern industry began to emerge, clusters formed organically as tradesmen grouped together to leverage economies of scale and to more effectively compete for business. A history of the original industrial revolution in Britain testifies to the importance of such clustering (Mathias 2001):

> Very shortly other 'external economies' developed. Once a pool of skilled labour grew up in a mill town that added to the 'inertia' of location. It made it more worth the while for expansion to occur in the same locality. A factory-trained labour force, of semi-skilled women and adolescents, was also an immense local advantage by the second generation. Another very important external economy was the convenience of specialized service industries—such as the bleaching firms, the machine-making shops, machine-servicing facilities which grew up in the shadow of the mills. All these things exercised a 'centripetal' pull on the cotton industry.

However, industrial clustering should be differentiated from simple agglomeration. While not a cut and dry proposition, one key difference is the degree of backward and forward linkages between firms (Karlsson et al. 2005). Some regions, perhaps because of easy access to a vital natural resource tend to specialize in the production of a particular good. While such groupings may contribute to certain positive externalities such as a deep talent pool, they may not on their own lead to an innovative or competitive environment (Delgado et al. 2010).

Local Living Conditions—Amenities as a Strategy for Talent Attraction

While the greatest force which pulls skilled workers to a cluster is the promise of continuous employment because of the large number of specialized local firms, secondary locational traits can help to lure employees towards an emerging cluster. Bangalore sits on a plateau, unlike other major Indian cities which are located near the ocean or in tropical lowlands. The pleasant climate is a real advantage. Boulder, Colorado, a fledgling tech hotspot, is located on the front range of the Colorado Rockies. The scenic views and opportunities for outdoor recreation represent a significant recruiting tool, as employers seek to attract highly educated and highly mobile workers. Universities, too, serve to enhance the appeal of an area. Cultural events such as concerts, lectures, and art exhibits that universities often sponsor provide opportunities for recreation and intellectual stimulation which may be otherwise lacking in industrial towns. Developing countries with significant foreign diasporas seek to attract their citizens back home with similar incentives. For top performers they offer high-quality housing, personal attendants, drivers, and recreational facilities along with plum administrative positions.

Linkages are crucial, especially between SMEs. One of the advantages of a large corporation is the degree of communication that can occur within a company. Bureaucratic politics aside, employees of the same large company are essentially working towards the same goal. But SMEs are often in direct competition with one another. Strong communication that leads to innovation separates an innovative cluster from a stagnant agglomeration.

Backward linkages are the connections between businesses and their suppliers. *Forward linkages* are the ties between businesses and their customers. The more information that flows up and downstream, the more innovative and responsive a company can be. Knowing that a battery supplier is close to a breakthrough in lightweight battery research and also having a market survey which shows that joggers dislike the heavy weight of current music players, could put a company in a good position to develop a new model music player developed specifically for the jogging market. Without the information the company might continue to produce the same heavy music player mindlessly until it was forced to adopt the new battery by its competition.

Cities: People Magnets in Flat World

Thomas Friedman (2005) popularized the concept of the "flat world" in which information and communication technologies combined with widespread political and economic reforms over the last 20 years have changed nature of international trade and competition. While previously nation-states and then multi-national corporations were the main drivers of globalization, Friedman argues that individuals are now competing on a global scale. Furthermore new technology means that the best and brightest from all over the world can compete without needing to move to a "leading" country to be successful.

Richard Florida (2008) also views people as the key to public policy surrounding innovation. In contrast to Friedman though, he argues that people's talents aren't likely to be fully expressed unless they can live in close contact with other skilled people. Florida looks to cities as the engines of economic growth, and says that while the world may be flattening for 2nd and 3rd tier cities and workers in manufacturing, 1st tier cities with a high degree of innovation are pulling even further ahead. He calls these cities "spiky" because of their high degree of economic and innovative activity in contrast with the surrounding countryside.

Florida points out that people look for different amenities in cities at different times during their lives. Young people are looking for lots of economic activity and a large potential mating pool. Middle-aged workers tend to want safe neighborhoods and excellent schools for their children. Top knowledge workers want to live in diverse cities that accept creative individuals and their sometimes non-conformist behavior.

5.2.4 Case Studies in Cluster Formation

5.2.4.1 Silicon Valley

Much of the enthusiasm for clusters is linked to success of the first, modern high-tech cluster, Silicon Valley. Despite advances in other regions throughout the US and the rest of the world, this area south of San Francisco, California still attracts the best and brightest minds in engineering, software, and web development. Silicon Valley did not emerge as the tech powerhouse it is today overnight. In fact, the San Francisco Bay region has been an important center for innovative radio and electronic research since the early twentieth Century.

Silicon Valley's name though, is a hint at the key driver of large scale growth. The development of the transistor or semiconductor, a key ingredient of which is silicon, was central to the region's success. The Dean of Engineering at Stanford University, Frederick Terman, helped create the Stanford Industrial Park in 1951. Companies, including many founded by Stanford grads, moved onto this real estate to be closer to the research being done at the University and to have better access to promising young engineers. Beginning with the seminal Shockley Semi-Conductor Laboratory in 1955, a series of spin-offs and startups led to rapid innovation in the high-tech electronics field. These early firms were heavily supported by procurement from the US government, especially the military which used the hardware in aircraft, missiles, and other advanced weaponry. Activity was accelerated by the spin-off culture. Partially a result of the region's existing business culture, it was also aided by the state of California's ban on non-compete contracts. In many states employees are barred from starting work on new projects that could directly compete with their former employer. In California, without such restrictions, there are stronger incentives to take advantage of business opportunities provided by technological advancement.

Technical expertise and an entrepreneurial culture weren't the only factors contributing to the Valley's rise. As early as the late 1960's, Venture Capital firms and boutique law firms began to do business in the area. These specialized legal and banking services made it easier for first time businessmen to make the leap from employee to owner. As the number of people with start-up experience grew, there were more opportunities for mentoring relationships to develop. Experienced investors guided their protégés in business development. Strong social links were formed between entrepreneurs, stimulating the flow of information about technological developments and investment opportunities.

Some of the drivers of Silicon Valley's growth have remained constant; a cooperative, collaborative, and entrepreneurial business climate, a strong talent base of scientists and engineers, regional pride and rivalry, and close university-industry relations. Others have developed later and aided growth or have faded away, such factors were; government procurement contracting, venture capital infrastructure, specialized legal firms, high intra and inter-national immigration, and cheap land values (Kenney 2000; Hospers et al. 2009).

5.2.4.2 Bangalore, India

Bangalore in the state of Karnataka, India was once known primarily as a resort for retired persons. Today it is the third most populous city in India and the center of the country's telecommunication, defense, computer, and IT industries. With a fast growing and dynamic economy, Bangalore attracts skilled engineers from across India and transnational corporations hoping to utilize this talented, skilled workforce at lower cost than in the West.

Bangalore's success stems in part from two structural components which are similar to Silicon Valley. The first is presence of large companies working for the Indian government working to develop high tech products for telecommunications and defense. The second is the large number of quality post-secondary educational institutions in Bangalore. The decision to concentrate such activities in Bangalore was made years ago when India maintained a highly regulated domestic economy. As trade liberalization began in the late 1980s and early 1990s, exposure to imported goods produced by foreign manufactures increased the level of competition among firms to produce higher quality products. Businesses owners in the region are tightly linked through a variety of ties, including college alumni and business clubs.

The opening of a Texas Instruments plant in Bangalore in 1985 was a watershed moment. Since then, many other foreign technology companies, including Google, Microsoft, IBM, and Oracle, have invested in Bangalore, often in one of two high-tech industrial parks, Electronic City and Whitefield. Many foreign companies view Bangalore as a cost effective location for research and development. Indian high-tech companies specializing in IT, engineering, and management consulting have seen rapid growth. Wipro and Infosys are the second and third largest ICT Indian ICT companies and are headquartered in Bangalore. From 1995–2005 the ICT sector has grown to over 70% of Bangalore's total exports. In 1995 Bangalore's ICT sector accounted for less than 0.25% of India's total exports, by 2005, that figure had reached 6%. Bangalore stands as a prime example of how to leverage its strengths: English speaking, high skilled, low cost labor to attract foreign companies and in turn foster the development of innovative and globally successful domestic firms (Van Dijk 2003; Grondeau 2007).

5.2.4.3   Silicon Wadi (Israel)

Over the past 20 years, Israel has established itself as a world leader in a variety of ICT businesses. This success stems from a variety of factors, including deliberate government policy. Israel's human capital provides its main competitive advantage. Israel's commitment to education, especially in computer science and engineering, along with an influx of scientists and engineers from the former Soviet Union in the early 90s, have provided a strong pool of potential knowledge workers. These workers have strong networks with one another because of the small number of Israeli universities and compulsory service in the Israeli Defense Force (IDF).

Israel spends a sizable portion of its budget on military R&D and in the 1960 and 1970s made significant advancements in secure networking and encryption technologies. This in-country research placed Israel in a strong position when the internet began to mature and a need for such technology became apparent. As new firms began to grow, a need for stronger venture capital markets was identified. In response, the Israeli government set up a special venture capital program called Yozma in 1993, which promised to match private investment in Israeli technology companies. Since then it has seeded 10 VC Funds with $20 million each giving them a 40% Government share and 60% private. Eventually, in all but one case of these seeded funds the government share was bought out by private investors. Today, total venture capital under management in Israel stands in excess of $10 Billion with around $1.5 billion invested annually (Wylie 2011; Engel and del-Palacio 2011). The Israeli government also started a number of incubators but after poor initial performance these were privatized and have since become more successful.

Like other developing clusters, Israel has successfully leveraged its nationals living abroad. Significantly, it has recruited Israeli engineers and entrepreneurs working in the US to develop strong links with Silicon Valley. A few years ago Silicon Wadi boasted the highest number of non-US companies listed on the NASDAQ exchange, while many American firms already operated subsidiaries within Israel (Bresnahan and Gambardella 2004).

5.2.5 Can Governments Stimulate Cluster Growth?

Every city planner, regional politician, and national economic official hopes to emulate the success of Silicon Valley or one of the other dynamic regional clusters mentioned above. But each example hints that "blank slate" innovative industrial development is not a simple, fast, or easy process. Various strategies have been used to stimulate "cluster-like" economic development across both the developed and developing world. The good news is that some policies can improve the performance of local firms and spur innovation. The bad news is that there is no "out of the bottle" solution for creating high-tech innovative clusters. Most cluster-based development policies have been at best mildly helpful. At worst they use up resources that could better be used elsewhere and produce no discernible impact (Braunerhjelm and Feldman 2006; Colombo and Delmastro 2002).

> **Korean Clustering—Grappling with Tradition**
>
> For the last half century of Korean economic development, young clever workers have sought corporate positions in the Chaebols (large conglomerates). These leading companies were considered national champions and employment at a chaebol carried great social prestige. Entrepreneurship was seen skeptically, an indication that someone had failed to make the cut at a larger firm. However, as the Korean government has recognized the economic potential of small, innovative startups (and the limits of older industrial policies), the authorities have taken steps to encourage dynamic technology clusters. One such example is DaedeokInnopolis located in Daejeon, Korea, south of Seoul. DaedeokInnopolis started as a science park called Daedeok Science Town in 1973.
>
> Despite having the advantage of being collocated with KAIST, Korea's leading research university, and significant government and corporate support, the science park was not particularly successful in stimulating the formation of new high-tech firms. The government has struggled to turn the science park into a self-sustaining cluster. Since the 2005 renaming of the science park, Daedeok has begun to see improved performance, between 2005 and 2009 sales increased from $2.5 to $12.3 Billion. Additionally it added 13 new companies to the KOSDAQ, an impressive number since previously the park had only produced 11 in total. However, the challenge of altering

Korea's traditional business culture will remain. Tax rules have been changed to allow new family businesses to enter the tax system more easily and bankruptcy laws have been altered to make the consequences of failure less dire (Watson 2011). The new Korean administration is pressing very much in that direction under the banner of the "creative economy".

5.3 Science Parks and Incubators

This section examines two related strategies for promoting innovation and regional economic development. *Science Parks* or *Research Parks* are mixed-use real-estate developments built close to Universities which seek to encourage Industry-University knowledge transfer. *Business Incubators* are also often located near universities (sometimes within science parks) and offer incentives such as low-rent property and networking opportunities to encourage spin-offs from university research and the establishment of new firms by entrepreneurs.

5.3.1 Science Parks

Taking Stanford's pioneering park as an example, many universities began building science parks and encouraging private industry to open branch research offices on or near campus where they would have easy access to talented graduates. The goal was increased knowledge spillovers and product commercialization. Science parks were envisioned as a location where government, industry, and the university could collaborate and share ideas. This collaboration would hopefully result in entrepreneurship and human capital development, which could serve as kernel for developing a regional agglomeration of knowledge workers.

Another impetus for creating science parks was desire to garner greater benefit from science research. In the United States, a great deal of public research funding is funneled through university departments. The rationale for basic research was partially predicated on the assumption that such research would lead to economic growth. As public science funding came under budget pressure in the 1970s and 1980s and as the US faced economic competition from Europe and Asia, science parks began to be seen as method for increasing technology transfer. Since the emergence of the first science parks in the United States during the 1950s, the concept has proliferated with over 400 parks worldwide. In North America there were 174 research parks by the middle of the previous decade which collectively employed over 350,000 workers and occupied over 47,000 acres (Battelle 2013) (Fig. 5.2).

At their start, science parks were essentially real-estate developments aimed at attracting high tech firms. Local municipalities or Universities used the prospect of cheap land and tax incentives to encourage high tech industry to move to the

Typical North American Science Park		
Size	Financing	Tenants
750 Employees 114 acres 6 buildings 314,400 sq. ft. of space, 95% occupied Only 30% of total estimated sq. ft. at build out currently developed 30,000 sq. ft. of incubator space	Less than $1 million per year operating budget Revenues primarily from park operations but funds also come from universities and state, local, and federal government Limited or no profitability; 75% of the parks have no retained earnings or retained earnings of less than 10% per year.	72% are for-profit companies 14% are university facilities 5% are governmental agencies Major industry sectors: IT, drugs and pharmaceuticals, and scientific and engineering service providers

Fig. 5.2 Science Park Characteristics (Battelle 2007)

research park. One of the primary reasons for the creation of science parks in the developed world has been the relative resiliency of universities in the face of economic decline. In many regions which have experienced de-industrialization, universities remain one of the few functioning large institutions and so attempts at economic rejuvenation are centered on the university. A similar logic prevails in developing countries, which are attempting to build an innovative environment from scratch. In either case, ties to a university lend credibility to such developments and imply a longer-term commitment by policy makers.

Ciudad Del Saber (City of Knowledge)—Redevelopment

As Panamanian officials prepared to take control of the former US-controlled region called the Canal Zone, they looked for ways to utilize the buildings and other infrastructure that were being abandoned by the Americans. Ciudad del Saber (CDS) was established by a private, non-profit organization in 1999, at the site of Fort Clayton, a former US military base. CDS houses a variety of affiliates within its properties including businesses, educational programs, and international organizations and NGOS. The park focuses on five major "work areas": Information Technology, Biosciences, Environmental Management, Human Development, and Business Management and Entrepreneurship. CDS also houses an onsite business incubator. Some of the main draws for the park are its reliable access to electricity and telecommunications, a business friendly tax policy, and proximity to Panama City and a nearby tropical ecosystem region called the Panama Canal Basin. CDS has become a UN hub (housing many UN agencies servicing Latin America), and currently houses 27 academic affiliates, 59 business affiliates, and 53 NGOs/IOs.

The sophistication of science parks has increased since their initial development. Initially, land and access to skilled graduates of the university were the main draws for business to move into the parks. As it became apparent that these loose ties were ineffective in promoting robust development, policy makers began to recommend a more activist approach to park administration. Stronger ties between faculty members and park tenants were encouraged. Business assistance services became more common. The focus began to shift from recruitment ties with large corporations to promoting the establishment of start-up companies. Efforts to increase the number of innovative small businesses led to the incorporation of business incubators into many science parks (Fig. 5.3).

Hsinchu Science and Industrial Park Taiwan—A Success Story

Beginning in the 1960s and 1970s, Taiwan began to be seen as a low cost manufacturing destination for basic electronics for foreign firms, and local SMEs began to make imitation products. In 1980 the Taiwanese government decided to invest in a science park near two well-regarded technical universities. Additionally, the organization exemplifying the national effort to research semi-conductors, ERSO (Electronics Research and Service Organization) was moved into the park. The park hosted many small emerging computer and electronics companies that were augmented by the government's policies for seeding venture capital funds. Rather than backing individual companies (picking winners and losers) the government sought to create a competitive local environment with incentives tilted towards the creation of IT companies.

One key to the park's success was luring back Taiwanese scientists and engineers who had been living and working in Silicon Valley. These individuals were offered substantial incentives, such as 49% government investment in any firm they started within the park and management positions within companies and park administration. These returnees brought with them knowledge about how to start and run high-tech companies and also founded the first private VC funds in Taiwan. The science park augmented the knowledge base of local companies which were already aggressively expanding. The park served to funnel knowledge from the universities and abroad into the private sector. From 1988 to the pre-recession height of 2007, annual sales from the science park grew 132% from 489.86 Billion NT to 1.14 Trillion NT (US $37 Billion) (Bresnahan and Gambardella 2004).

Proximity between industry and universities does not automatically result in collaboration. Science parks may succeed on some level, but there is little hard empirical evidence to suggest they stimulate new economic growth. They do provide an environment conducive to communication and coordination between industry, government, and academia. The most effective parks are deeply integrated into the communities where they are located. They

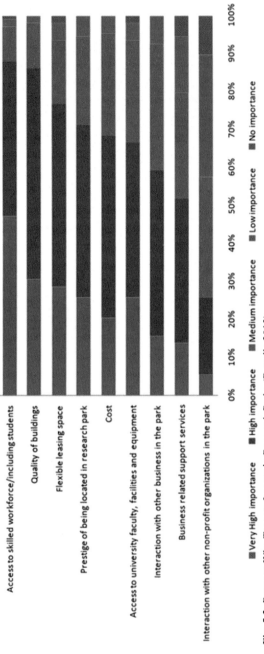

Fig. 5.3 Reasons Why Tenants Locate in Research Parks (Battelle 2013)

must acknowledge the occasionally competing goals of various local stakeholder groups. These goals include providing jobs for local workers, corporate access to university R&D, regional development, and enhancing university prestige and revenue from technology transfer. Policy makers must also realize their own bias looking at "success stories".

In certain cases, universities and their industrial relations have played a key role in producing self-sustaining clusters, but many other factors are responsible for regional economic development. Realistic time horizons must also be kept in mind. Even successful science parks such as Research Triangle in North Carolina have taken over 50 years to become fully established. An attractive campus with several prestigious sounding businesses grouped closely together may make science parks an attractive option for policy makers seeking an impressive looking end product, but they are unlikely to rapidly contribute to economic growth and development (Bresnahan and Gambardella 2004).

5.3.2   Knowledge Business Incubators

Widely used by local governments to encourage general entrepreneurship, business incubators which specifically focus on high-tech sectors are a sort of inversion of the science park model. Whereas science parks try to attract businesses to co-locate and hopefully collaborate with universities, business incubators seek to encourage spin-offs and start-ups. Incubators try to create a welcoming environment for entrepreneurship by lowering startup costs and providing consulting services. Key features of incubators are temporary leases in business rental property offered at below market rates, professional business managers, and structured networking opportunities with venture capitalists.

Technological MIDI—Brazil

The southern Brazilian city of Florianopolis has sought to encourage the development of a high-tech innovative economy but faces difficulty because of its distance from the commercial and financial hubs of São Paulo and Rio de Janeiro. The local technology business council ACATE, founded Technological MIDI in 1998 with the aim of incubating up to 10 companies. In 2001 they expanded the facility to house a total of 14 companies. MIDI offers many of the same services as other incubators including rent at half the market rate, access to business and financial networks, business consulting, and tax relief. Its close ties to the local business community and the national government are helpful as well. It is registered to receive federal subsidies under Brazil's so-called "IT Law", which encourages domestic IT innovation. By 2007, companies which had graduated from the incubator had achieved sales of US $13.9 Million and employed 385 people. This success earned the incubator the best technology incubator award in 2008 from the Brazilian innovation and entrepreneurship association ANPROTEC.

Some of the key services provided by business incubators include (Johnsrud et al. 2003):

- Provision of a facility to house client firms, including office space, business services and access to laboratory and other technical resources needed for prototyping, testing and analysis for technology-based clients
- Agreement among stakeholders on the objectives of the incubator, including short-term and long-term expectations about tenants' growth and maturation
- Experienced incubator managers who can design and deliver customized services to address the unique needs of client firms
- Design or use of long-term financial support strategies that draw on locally available investment sources, client fees, and downstream equity or royalty returns
- Reliance upon a supportive community infrastructure to facilitate access to the widest possible range of financial, management, marketing, technical, legal and information resources needed for tenant training, networking, market analyses, regulatory compliance and product development.

Business incubators have become even more widespread than science parks, not least because of the fewer resources needed to establish one. Incubators carry additional appeal because of how far along they are in the continuum from basic research to marketable product. The primary rationale for high-tech business incubators is that small, innovative companies are the most likely to create transformative technologies that will benefit society at large, potentially even leading to the creation of new industries especially in advanced economies.

Beginning entrepreneurs have difficulty evaluating the market potential of innovative technology, and even less understanding of the necessary steps towards commercializing a product. This experience gap is a serious barrier to universities that are encouraging their faculty members to spin off new firms. Since such businesses are inherently risky and unproven, they suffer from a lack of investment. Governments seek to correct for this market failure by subsidizing the establishment of such firms (OECD 2006). Incubators attempt to bridge this gap in three key ways, by providing infrastructure, business support, and access to networks.

5.3.2.1 Infrastructure

New businesses face substantial hurdles in acquiring office space, support staff, parking, storage, telecommunications, and other basic overhead requirements. Business incubators help new businesses by simplifying this tedious and time-consuming phase of establishment. By offering package deals at below market rates, firms find themselves at an immediate advantage. The act of renting a real office (rather than maintaining a virtual office or working out of a home) confers added legitimacy to new firms at time when this image is especially important for attracting investment. Business incubators typically house multiple firms. These firms are able to share the costs of the various services such as a receptionist, audio/visual equipment, printers and faxes, and insurance. Interactions between tenants can stimulate further growth as synergies between complementary firms can develop.

5.3.2.2   Business Support

The level of business support varies widely by incubator but can include help with composing business plans, mentoring or coaching from more experienced managers, training sessions, accounting services, IT support, legal assistance, as well as other options.

Equity Stakes

Incubators that are started by the public sector tend to be focused on economic development and increasing local employment levels. As a consequence they tend to ask little financially in return from the entrepreneurs they host. Private incubators, however, may require an equity stake from the firms they incubate. Leading private sector incubators, Y-Combinator and Techstars require around a 6% equity stake from their startups. This cuts both ways; entrepreneurs trade away some of their value, but this incentivizes incubators to work harder since they will share in the final success of the start-ups. According to the National Business Incubators Association, 24% of US tech incubators require some sort of equity stake (Bass 2012).

5.3.2.3   Access to Networks

Of course the main barrier start up business face as they attempt to expand is access to capital. Often this is because entrepreneurs lack contact with venture capitalists and angel investors. Incubators can arrange for their tenants to network with prospective investors. This could be in the form of events where founders make pitches to investors or business lunches with local leaders to gain social capital.

Other Types of Incubators

Accelerators: Rather than allowing for slow growth like more traditional incubators, business accelerators aim to rapidly bring entrepreneurs from the initial idea phase forward to a solid business plan and a prototype or website. They often try to connect budding firms to venture capitalists or angel investors. Examples include Y-Combinator and TechStars.

Virtual Incubator: Some firms already have office space or infrastructure in place but need help with other aspects of business development. Virtual incubators use the internet to connect entrepreneurs with management counseling and other services without having to move to a central shared location.

5.3.2.4 Assessments of Effectiveness

The word "incubator" is key. Business incubators aim to "graduate" companies from the incubator and into the regular market once they become established. Successful operation of a knowledge business incubator requires solid selection criteria and robust standards for firm exit. By being selective about which firms they choose to house, incubators can increase the chances of success. Similarly being clear about when firms must exit provides certainty and encourages firms to expand quickly to become profitable enough to survive outside the incubator.

ICEHOUSE—New Zealand

The ICEHOUSE business growth program was founded in 2001 as part of an effort to increase the number of high-tech SMEs in New Zealand. Its stated purpose was to launch 350 firms to meet a national goal of 3000 new SMEs. Partners for ICEHOUSE included the University of Auckland Business School, BNZ, HP, NZTE, Gen-i, Ernst & Young, Paul Diver, Grafton Consulting Group, and Microsoft. The ICEHOUSE incubator is linked to New Zealand's largest network of angel investors and has a monthly event where 25 entrepreneurs can attend a seminar explaining how to launch a start-up and have the opportunity to meet with incubator staff about joining ICEHOUSE. The incubator offers 3 basic levels of service, market validation, business plan development, and full incubation. ICEHOUSE primarily aims to incubate companies with an intellectual property component with a high growth potential in an emerging market. Since 2001 it has launched 75 companies and attracted $50 Million in angel investment. It was ranked as one of the top 10 technology incubators in the world by Forbes magazine in 2010.

Y-Combinator

One of the most successful tech accelerators was founded in 2005. Y-Combinator takes selects prospective entrepreneurs through 3-month "bootcamps" designed to quickly launch promising companies. After the 3 month period, entrepreneurs pitch their ideas to a group of investors and venture capitalists at a presentation called "Demo Day", an event which has become extremely influential.

Applicants are rigorously screened, but Y-Combinator seeks to invest a small amount of money across a large number of companies. Those accepted into the program are given approximately $18,000, business training, and access to Y-Combinator's network of experienced entrepreneurs and investors. The budding companies cede a 6% equity share to Y-Combinator in exchange for their service. To date, this accelerator has launched 380 companies, including notable internet businesses such as Dropbox and Reddit. (http://ycombinator.com/2012)

It is critical for policy makers and the managers of a business incubator to be clear about the objective of the operation. There is a wide range of potential incubator sponsors including municipalities, universities, government agencies, and non-profit agencies, who may all seek different end goals. This includes privately owned business incubators whose goal is to achieve a profit. This can sometimes come at the expense of local economic development. If the goal of a knowledge business incubator is to spawn numerous high technology companies to stimulate growth, this must be explicit. Managers may choose to allow successful businesses to remain on site too long because of the steady revenue from rent and fees. Similarly, they may allow unrelated businesses to rent what is essentially subsidized office space. These practices consume resources that could be used by desired technology startups.

Knowledge business incubators are a cost-effective way of stimulating the creation of high tech businesses and fostering a local culture of innovation. However they must be carefully managed and focused on their specific objectives. Innovation and quality should be more highly prized than simply filling space within the incubator (Lalkaka 2002; Almubartaki et al. 2010).

5.4   Conclusion

The process of creating self-sustaining high-tech clusters cannot be fully controlled by governments. While there is no well-defined recipe for this type of economic development, certain ingredients may be helpful. These include strong links between research and development at universities and emerging industries, access to capital markets, and a local culture of competition and collaboration.

Policies such as the creation of science parks and knowledge business incubators can help foster technology transfer and entrepreneurship but are unlikely to stimulate self-sustained economic growth in the absence of other factors. They cannot alone make up for deficiencies in local systems of innovation. Without stable macroeconomic environments, strong labor and capital markets, respectable intellectual property protection, a reasonable research and development base, rule of law, and other basic requirements, an entrepreneurial high tech business culture is unlikely to take hold (Fagerberg et al. 2005; Asheim et al. 2006).

References

Al-Mubaraki, H., Al-Karaghouli, W., & Busler, M. (2010). The creation of business incubators in supporting economic development. European, Mediterranean, & Middle Eastern Conference on Information Systems 2010 (EMCIS 2010), April 12–13, 2010, Abu Dhabi.

Andersson, T., Serger, S. S., Sorvik, J., & Hanson, E. W. (2004). *The cluster policies whitebook*. Malmo: International Organization for Knowledge Economy and Enterprise Development (IKED).

Asheim, B., Cooke, P., & Martin, R. (Eds.). (2006). *Clusters and regional development: Critical reflections and explorations*. New York: Routledge.

Bass, D. (2012). Bringing up business: Incubators through the eyes of the entrepreneur. Consumer Media Network. http://www.cmn.com/2012/05/bringing-up-business-incubators-impact-seen-through-the-eyes-of-the-entrepreneur/2012.

Battelle. (2007). *Characteristics and trends in North American research parks: 21st century directions*. Columbus: Battelle Technology Partnership Practice.

Battelle. (2013). Driving regional innovation and growth: The 2012 survey of North American university research parks. Columbus: Battelle Technology Partnership Practice. http://www.aurp.net/battelle-report. Accessed Feb 9 2014.

Braunerhjelm, P., & Feldman, M. (Eds.). (2006). *Cluster genesis*. New York: Oxford University Press.

Breschi, S., & Malerba, F. (Eds.). (2005). *Clusters, networks, and innovation*. New York: Oxford University Press.

Bresnahan, T., & Gambardella, A. (Eds.). (2004). *Building high-tech clusters: Silicon valley and beyond*. New York: Cambridge University Press.

Colombo, M., & Delmastro, M. (2002). How effective are technology incubators? Evidence from Italy. *Research Policy, 31*(7), 1103–1122.

Delgado, M., Porter, M., & Stern, S. (2010). Clusters and entrepreneurship. *Journal of Economic Geography, 10*(4), 495–518.

Engel, J., & del-Palacio, I. (2011). The case of Israel and silicon valley. *California Management Review, 53*(2), 27–49.

Fagerberg, J., Mowery, D., & Nelson, R. (Eds.). (2005). *The oxford handbook of innovation*. New York: Oxford University Press.

Florida, R. (2008). *Who's your city?* New York: Basic Books.

Friedman, T. L. (2005). *The world is flat*. New York: Farrar, Straus, and Giroux.

Grondeau, A. (2007). Formation and emergence of ICT clusters in India: The case of Bangalore and Hyderabad. *GeoJournal, 63*, 31–40.

Hospers, G., Desrochers, P., & Sautet, F. (2009). The next silicon valley? On the relationship between geographical clustering and public policy. *International Entrepreneurship and Management Journal, 5*, 285–299.

Johnsrud, C., Theis, R., Bezerra, M. (2003). Business incubation: Emerging trends for profitability and economic development in the US, Central Asia, and the Middle East. US Department of Commerce Technology Administration.

Karlsson, C., Johansson, B., & Stough, R. R. (Eds.). (2005). *Industrial clusters and inter-firm networks*. Northampton: Edward Elgar Publishing Limited.

Kenney, M. (Ed.). (2000). *Understanding silicon valley: Anatomy of an entrepreneurial region*. Palo Alto: Stanford University Press.

Lalkaka, R. (2002). Technology business incubators to help build an innovation-based economy. *Journal of Change Management, 3*(2), 167–176.

Mathias, P. (2001). *The first industrial nation: The economic history of Britain 1700–1914* (3rd ed.). New York: Routledge (1969).

Nadvi, K. (1995). Industrial clusters and networks: Case studies of SME growth and innovation. UNIDO SME Programme.

OECD (Organization for Economic Co-Operation and Development. (2006). The SME financing gap. *Theory and evidence*. Paris: OECD.

Porter, M. (1998). *On competition*. Cambridge: Harvard University Press.

Van Dijk M. P. (2003). Government polices with respect to an information technology cluster in Bangalore, India. *The European Journal of Developmental Research, 15*(2), 93–108.

Watson, J. (2011). *Fostering innovation-led clusters: A review of leading global practices*. London: The Economist Intelligence Unit.

Wylie, C. (2011). Vision in venture: Israel's high-tech incubator program. *Cell cycle* (*Georgetown, Tex.*), *10*(6), 855–858.

Y-Combinator. http://ycombinator.com/2012

Chapter 6
High-Risk Finance

Daniel Waggoner

6.1 Introduction

While government-funded research is an important component in an innovative economy, particularly at the more basic levels, most innovation in advanced economies is funded by the private sector. Within the private sector, a large portion of research is conducted by established companies, while a smaller but arguably more innovative amount of research is conducted by small, knowledge-intensive companies. This Chapter focuses on the challenges of financing a subgroup of these companies, knowledge-intensive startups, which have historically been a significant source of new innovations and job growth.

Knowledge-intensive startups begin with an entrepreneur willing to take a risk on starting a company to develop an idea that he or she believes brings something new and original to the marketplace. These startups are responsible for an outsized share of innovation in developed economies compared to their small size and relative share of research and the economy as a whole. Though they have the potential to be highly successful, they also have a high rate of failure. It is for this reason that they are considered high-risk investments. Many of them will fail, in fact, but successful survivors can make high returns for their investors. They also create substantial benefits for the economy and the public. In the U.S., startups have been responsible for virtually all new job creation over the past 30 years (Kane 2010).

Adapted from a chapter of the Innovation Policy Handbook report composed for the World Bank (2012). Original unpublished and available upon request.

D. Waggoner (✉)
Center for International Science and Technology Policy, The George Washington University, Washington, D.C., USA
e-mail: danny@dannymail.net

© The Editor(s) 2015   85
N. S. Vonortas et al. (eds.), *Innovation Policy,* SpringerBriefs in Entrepreneurship and Innovation, DOI 10.1007/978-1-4939-2233-8_6

High-risk financing bridges the gap between an individual with a great idea and a viable company with a new product on the market. Often, those with new and innovative ideas do not have the resources to develop a new product on their own, including both funds and the necessary set of business and management skills. This is especially true in sectors that require highly specialized knowledge and resource-intensive development such as biotech or information technology. High-risk financing fulfills a good part of this need and makes innovation by individual entrepreneurs and small businesses possible. This chapter first discusses how small startup firms are financed in developed economies and then examines the specific challenges to promoting high-risk financing in emerging markets.

6.2 Types of Financing

6.2.1 Debt and Equity

There are two main types of financing that are typically available to startup companies. The first is a loan, where a bank (or another lender) provides financing for a given term that must eventually be paid back by a business with interest. The other type is equity financing where an investor provides money for a startup in exchange for shares of the company. Often this equity-based financing is invested and managed by angel investors or venture capital firms, which specialize in developing small innovative startup companies. There are both benefits and drawbacks for each type of financing, which means that a startup company should carefully assess its circumstances before choosing a type of financing. Banks do not tend to concern themselves with the day-to-day management decisions of companies; their main interest is making sure that the loan is repaid on time. This preserves the autonomy of the management of a company, but it also has the potential to drain a company of financial resources, which must go toward loan payments, diverting capital from being reinvested into the new company (Ben-Ari and Vonortas 2007). During the early stages of new innovative companies, which usually have few or no sources of revenue and require large initial capital investments to develop their products, loan payments have a high opportunity cost. For this reason, loans are usually most appropriate for companies that already have steady revenue streams. For startups that are in their nascent stages and are focused on developing their first products, equity investments often tend to be the preferred option.

Equity investments usually place less financial strain on a business during its early stages; however, they often come with greater strings on management decisions (Hall and Lerner 2009). Investors, through acquiring partial ownership of a company, can place restrictions on management decisions. The interests of entrepreneurs and the investors may not always converge. Entrepreneurs may choose to pursue strategies that bring in more funding, such as issuing shares to more investors, while current investors may wish to limit those strategies in order to maintain the value of their current investment. This divergence has the potential to create friction between entrepreneurs and investors. However, input and direction from investors can be helpful to a new company. The management of a small startup company may consist of entrepreneurs who are more competent in an area of technical expertise than business management. For these individuals, guidance from investors can be very important to the success of the company.

6.2.2　Equity Investors Provide Useful Expertise

Though entrepreneurs may not enjoy relinquishing control to investors, venture capitalists provide managerial expertise that, in general, improves the performance of startups (Hall and Lerner 2009). Most venture capital funds employ compensation schemes for their managers that depend heavily on the performance of the fund's portfolio. This encourages fund managers to strictly monitor progress at a new company, particularly in the early stages when a company has developed very few assets that can be liquidated in a bankruptcy. However, some studies have not found a correlation between incentive pay and fund performance (Gompers and Lerner 1999).

The primary method of control employed by investors is releasing funds in short stages. If performance is poor or a venture capital company wants to force a change in a company that it has invested in, it will withhold funding. Venture capital firms will usually grant greater autonomy to companies that are performing well and place underperforming companies on a tighter leash by providing small installments of funding or even by taking over the management of the company. The funding duration is also usually very short when a company is new and its assets are intangible, such as knowledge retained by employees. Once an asset becomes more tangible, for instance through the acquisition of a patent, venture capital companies will lengthen funding cycles (Gompers 1995). When assets are tangible, the "salvage value" of a company increases since the assets can be sold if a company fails. This, in turn, decreases the financial risk to investors.

This "hands on" approach that most venture capitalists adopt is one of the reasons that venture capital tends to outperform other types of funding. Venture capital, according to some studies, encourages greater innovation other types of investment. One study found that venture capital funding will create three times as much patenting activity as an equivalent amount of corporate R&D (Kortum and Lerner 2000). Innovative firms that receive venture capital also tend to bring products to market more quickly (Hellman and Puri 2000).

6.2.3   Investors Hindered by Information Asymmetries

The differing levels of knowledge between entrepreneurs and investors, otherwise known as information asymmetries, can also create tension and raise the price of capital. Entrepreneurs tend to know their product (or potential product) better. As they are intimately involved with the development process of the product, they have a better grasp of the timeline for a completed product and whether it will be viable on the market. Investors usually have less technical expertise in the field than the entrepreneur, which creates an information asymmetry or "trust gap".

This trust gap can create communication challenges in the relationship between investors and entrepreneurs, as entrepreneurs attempt to keep their investors informed about progress in product development and the timeline for return on their investment. Investors may question the quality of the information since entrepreneurs have an interest in preserving funding sources and may bias information for their own benefit.

This trust gap also makes initial investment decisions more difficult for investors. Since entrepreneurs have a better understanding of the potential product and the probability of success, investors will be at a disadvantage. Providing too much information to investors also poses risks to entrepreneurs. Disclosing a new idea in too much detail may allow others to steal or copy it. The low quality of information available to investors during investment decisions increases the risk of investing in a dead end project and therefore will raise the cost of capital for innovative companies across the board (Hall and Lerner 2009).

These information asymmetries and high failure rates are more pronounced in innovative startups compared to new businesses in more established fields. This raises the price of external capital for innovative companies over what other startups would pay. For this reason, the cheapest option for funding R&D is using internally sourced capital, such as existing revenue streams or retained revenue (Hall and Lerner 2009). New startups do not have this option, so they must use the more expensive externally-sourced capital. This in turn places greater demands on future performance in order to pay for the more expensive capital.

6.3   Stages of Investment

It is rare to find one investor who will fund a new startup from beginning to end. Some corporations do this internally with new startups that are wholly-owned by the corporation, but in the case of individual entrepreneurs and small startups, new funding comes in stages. Entrepreneurs are also not limited to only one type of funding. Some companies will mix and match equity and debt-based financing at different stages of the company's development depending on the needs of the company at the time (Ben-Ari and Vonortas 2007). Each step in developing a new product—from idea, to research, to prototype, to a marketable product—requires larger amounts of capital. Advances in product development must be synchronized with new infusions of capital. When this does not happen, funding gaps threaten the survival of startup companies.

The initial funds for the very early stages of developing a concept into a business (known as the seed stage) will likely come from an individual's own finances or from a group of closely related people.[1] This amount of money is variable depending on personal wealth, and in most cases in the U.S., does not exceed a few hundred thousand dollars. To get beyond the seed stage, entrepreneurs require outside investors that are willing to make small investments in volatile, early-stage companies. In most cases, these investments come from wealthy individuals known as angel investors who invest a small percentage of their wealth in high-risk ventures. Should the company prove successful, some angel investors will continue to fund the company into the post-seed startup stage. Once funding requirements reach into the $ 1–2 million stage, the investments start to become large enough to attract the attention of venture capital funds.

Venture capital funds are different from angel investors in that they are not personal investments; rather they are most often limited liability corporations where the money from investors is controlled by professional managers. Venture capital funds have been gradually shifting their focus, investing in less risky, later-stage companies that require higher amounts of funding (PWC/NVCA 2014). In 1995, investments in seed companies represented 16 % of venture capital investments. By 2013, this number had shrunk to 3 % (PWC/NVCA 2014). Part of this could be due to the higher management costs of monitoring multiple small investments versus fewer larger investments. Another reason could be that venture capital funds have become risk averse (Fig 6.1).

Stage of Development Definitions

- Seed/Start-Up Stage
  - The initial stage. The company has a concept or product under development, but but has not fully geared up research or other operations. Usually in existence less than 18 months.
- Early Stage
  - The company has a product or service in testing or pilot production. In some cases, the product may be commercially available. It may or may not be generating revenues. Usually in business less than three years.
- Expansion Stage
  - Product or service is in production and commercially available. The company demonstrates significant revenue growth, but may or may not be showing a profit. Usually in business more than three years.

---

[1] Known as the three F's: friends, family, and fools

- Later Stage
  - Product or service is widely available. Company is generating on-going revenue; probably has positive cash flow. More likely to be, but not necessarily profitable. May include spin-offs of operating divisions of established private companies.

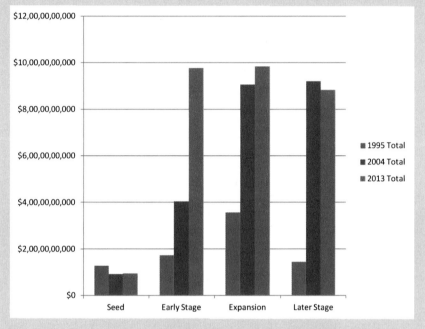

Fig. 6.1 VC Funding by stage of development. (Source: PWC/NVCA 2014, PWC 2014)

The change in the preference of venture capital funds has created a gap between the seed stage and startup stage, leaving fewer funding options for companies attempting to break through the seed stage. Angel investors, perhaps in an attempt to help plug this gap, have gradually shifted into post seed funding. In 2002, 47 % of angel investments went to seed companies, while 33 % went to post seed companies (Sohl 2003). By 2010, this had shifted substantially: 31 % of angel investments went to seed companies, while 675 went to post seed companies (Sohl 2011). This trend has also been observed in Europe, where seed funds now make up a small fraction of venture capital fund investments (EVCA 2010). Whatever the cause of this shift, it has left seed and early-stage companies with less opportunity to receive external funding (Table 6.1).

Table 6.1 Contrasting angel and venture capital Investors in the U.S.

	Stage of investment	Total number of deals	Total value of deals	Number of investors/funds	Percent of exits that break even or better
Angel investors[a]	31% Seed/67% post-seed	61,900	$ 20.1 Billion	265,400	50%
Venture capital[b]	3% Seed/68% expansion and later stage	3543	$ 23.3 Billion	842	80%

[a] Jeffrey Sohl, "The Angel Investor Market in 2010: A Market on the Rebound", Center for Venture Research, April 12, 2011
[b] "NCVA Yearbook 2012", National Venture Capital Association, April 2012

6.4 Exiting

The ability to recoup investment is of crucial importance to investors. The goal of an equity investor is *not* to become a business owner, but to buy shares in a company with potential, support it and help it grow, and then profit off of the investment by selling shares after company value has increased. Equity investors tend to be patient, but they will eventually want to sell their equity to realize their profits, which are often rolled over into new investments.

6.5 Acquisitions vs. Initial Public Offerings

An exit from an equity investment can take shape in three forms: a sale (or takeover by another company), an initial public offering (IPO), or bankruptcy. Selling the startup to a larger company or another group of investors is one of the most straightforward ways to recoup investment. Often small investors such as angel investors take a company through the seed stage and then sell their equity to a larger investor that can provide additional funds to continue to grow the company. A larger company may also purchase a startup for its intellectual property or product line. A purchasing company might want to buy a patent to improve an existing product or to prevent a competing product from coming to market.

Inventive Private Sector Solutions for Risk-Financing in the UK

While government support is often essential for overcoming funding gaps where there are high risks that discourage private investment, new investment methods for private capital may play a part in making those funding gaps easier to overcome. In response to the global credit crunch, several UK firms are experimenting with ways to provide capital to new businesses. A business called Crowdcube is banking on a concept called crowdfunding, where an entrepreneur presents a business plan to a website with thousands of potential investors and each investor is able to review the business plan and decide how much he or she wants to invest. This results in a large number of people each making small investments in a new business. This is in contrast to the traditional angel or venture finance model where a dedicated team of a few individuals reviews a new business and makes a large investment. By allowing individual investors to review the business plan on their own time, it avoids the lengthy and costly review process that takes place with traditional equity finance. It also allows a greater number of individuals that might not have larges sums of money available to make small investments in new businesses and thereby spreads the risk for potential failures. Individual investors may not have the skill of dedicated angel and venture investors in reviewing business plans and picking successful winners; however, providing a way for them to participate in financing a new business may bring significant amounts of new capital into play.

Another UK company focuses on providing crowdfunding through debt rather than equity. Instead of lots of investors making small equity investments in different businesses, Funding Circle's model allows lots of lenders to make small loans to new businesses. Funding Circle first reviews the risks involved for a potential borrower and then allows its members to provide loans through an auction. Each potential lender bids an interest rate at which he or she is willing to provide a loan, and then the lender with the lowest interest rate will be selected to make the loan. Just as in CrowdCube's model, Funding Circle encourages lenders to make at least 20 small loans to spread the risk around.

Both of these companies use a crowdfunding model that provides a new set of benefits and risks to the marketplace. By democratizing finance, more people with less means will be able to invest and potentially profit from new startups. However, most investors will be at a greater risk for losses since they will not have the experience to evaluate what businesses might be successful and the information available on a website may not be as good as what angel and venture investors receive when they invest in a company. This is an inventive way for investing that, if it proves to be successful, may open new doors for both private investors and entrepreneurs.

While acquisitions are common, estimating the value of a privately held company is difficult for a purchaser. In contrast to an initial public offering on a stock market, the purchaser cannot benefit from a market valuation of the company. The buyer must resort to other methods such as making comparisons to similar publicly traded companies. Stakes in private companies are also less liquid than publicly traded companies, making it more difficult for purchasers to sell their equity. The risk associated with purchasing a private company that is illiquid and difficult to value results in a sale price discount that is often *ad hoc* (Das et al. 2003). The benefit of exiting through an acquisition is that it avoids the costly and administratively cumbersome requirements to which publicly traded companies must adhere. Prior

Table 6.2 Exits of venture-backed companies in 2011

	Number of deals/ offers	Total value of deals/ offers	Average size of deals/ offers
M&A [a]	467	$ 24.08 Billion	$ 145 Million
IPOs	53	$ 9.92 Billion	$ 182 Million

[a] Deal amounts were only made public for 166 out of the 467 total deals. These numbers only reflect those deals for which data was disclosed

to an IPO, a company must list itself on a stock exchange and comply with the requirements of the particular exchange, such as minimums for market capitalization, share issuance, and historical earnings. A publicly-traded company must also comply with public reporting requirements and accounting standards that vary from country to country and are typically much more burdensome than the requirements for privately-held companies. The added burdens of IPOs mean that private sales are often the only feasible option for small and early stage startups (Table 6.2[2]).

An initial public offering allows owners to tap funds from a large number of investors. During an IPO, a company will list itself on a stock market and make a number of shares available to the public for purchase. This allows a company to raise a large amount of funds quickly which do not have to be repaid to the investors purchasing the shares; however, it subjects a company to a number of constraints faced by publicly traded companies, such as complex accounting and reporting requirements, shareholder relations, and greater information disclosure (as discussed above). Despite this, an IPO is the preferred method of exit for a venture capital fund because it usually yields the highest return on investment. Unlike a takeover or private sale, where shares and ownership are transferred immediately, venture capital funds retain their shares for an average of a year after the IPO (Megginson and Weiss 1991). An immediate sale during the IPO would signal that the asking price for the stock is too high and that the company is overvalued. Venture capital funds hold on to their shares and their board positions to signal stability in management and confidence in the stock.

In contrast, IPOs are rarely used as a means of exit by angel investors. In 2010, IPOs accounted for less than 7% of exits for angel investors. The majority of exits, 66%, were through mergers and acquisitions (private sales) while the remaining 27% were divested through bankruptcy (Sohl 2011). The early stage focus of angel investors means that companies are usually not mature enough for a public offering when an angel investor is ready to exit. Private sales are also better suited for early stage companies because the intangible intellectual assets are not likely to be well understood or valued by the public. Experienced investors, such as venture capital funds, will more likely provide greater value in a private sale as they have a higher capacity to understand the value of the knowledge-based assets.

---

[2] National Venture Capital Association and Thomson Reuters."Venture Backed IPOs Have Strongest Opening Quarter in Five Years." Press Release. April 2, 2012. Available at: http://www.nvca.org/index.php?option=com_content&view=article&id=344&Itemid=103

6.5.1 Bankruptcy

Bankruptcy is the least desirable option for exit. While bankruptcy can be viewed as a failure, it does not represent a total loss for investors. Most companies have some salvage value that will allow investors to recoup a portion of their investment when the assets are liquidated. For professional investors, bankruptcies happen in a minority of cases. Even following the financial crisis when business bankruptcies peaked in the US in 2009, bankruptcies remained in the minority, accounting for 40% of investment exits for angel investors (Sohl 2010). For small businesses in general, around 70% survive the first two years and 50% survive for five years (SBA 2011). On average, the percentage of business angel investments that have ended in bankruptcy has remained below the average for small businesses in general. This indicates that there may be two factors that distinguish success rates for angel investor-backed companies from that of small businesses in general. The first is that angel investors are more likely to invest in businesses that are more likely to succeed. The second is that angel investors provide valuable advice and direction that help businesses succeed.

6.5.2 The Cost of Failure Matters

The process of creating innovations entails trying new things that have unknown outcomes. There is inherently higher risk in this process, making some bankruptcies unavoidable. This can discourage entrepreneurs from taking a chance in the first place. Making it possible for people to recover from a failed business will help mitigate this risk. The easier it is to obtain funding for a new project after a failure, the more likely an entrepreneur will abandon a project that has a lower chance of success (Landier 2006). If decreased access to future capital makes failure too costly, entrepreneurs will tend to commit themselves to lower performing projects for longer periods of time. Though this tends to decrease the failure rate of firms, it also diverts capital toward lower performing projects and away from those with the highest chances of success. The consequences of failure must adequately discourage reckless use of funds without discouraging risk-taking altogether.

6.5.3 Ease of Exit

The ease of recouping investment is a very important factor in determining the level of venture capital investment in a country (Black and Gibson 1998). A key reason for the high level of venture capital funding per GDP in the US is its robust IPO market (Black and Gilson 1998). A well-developed IPO market provides the best means for extracting as much profit as possible from an equity investment. This provides investors with a greater degree of confidence that they will be able to profit on their investment once they decide it is time to sell. Venture capital tends to lag behind in emerging markets due to the lack of an established equity market and therefore a market for IPOs. They also in general lack the liquidity through which such a market could be developed, largely due to the lack of pension fund investments (Mani and Bartzokas 2002). Developing an equity market is a large challenge for an emerging market. This leaves room for some innovative solutions from policymakers and business leaders that will provide alternative avenues for venture capital funds to maximize the return on their investments once they decide to exit.

6.6   Contextual Challenges

Investors do not like risk. They despise uncertainty. Investors only choose to take on risk if they believe the potential reward outweighs that risk. The technology and market risks they undertake in a knowledge-intensive startup are already quite significant but they consider them to be within their sphere of influence. Any factor that increases risk beyond that level, especially where it takes on elements of political or economic uncertainty, discourages investment if it is not counterbalanced by an increase in potential return. In any market, risk does not lie in the quality and viability of an innovation alone, but also in environmental conditions.

One risk that any market faces is cyclical market downturns. A company that would normally thrive in a booming market might need to shut its doors during a downturn if demand for its product dries up or capital flees for safe haven investments. This risk is systematic and there is little that investors can do about it. They will try to manage unsystematic risks like those detailed below.

6.6.1   Intellectual Property Rights

Intellectual property protection is critical for innovative, knowledge-based firms. A company must be able to exclude others from using its research and inventions or else it will not be able to capture all of the value from its intellectual property. When a company is unable to exclude others, most commonly accomplished by enforcing a patent, another company can profit from using an invention without bearing any of the costs of the invention's development. The creator of the invention then becomes an unwilling subsidizer of a competitor. A country must have enforceable intellectual property rights to protect the money and resources a company invests in research and development, or else investors in the company will not want to run the risk of having a knowledge-intensive company's most valuable asset, its intellectual property, pilfered by a competitor. This both devalues the assets of knowledge-intensive companies and lowers the potential for profits. Weak property rights increase risk and discourage investment.

6.6.2 Taxation

Capital gains taxes directly affect the rate of return that equity investors achieve from their investments (European Commission 2002). Targeted breaks in capital gains taxes for angel and venture financing can increase returns and help offset the costs of the increased risk of investing in early stage companies. The United States saw a meaningful increase in early stage angel investments after it introduced temporary but steep cuts in capital gains taxes for individuals investing in certain types of companies with less than $ 50 million in assets (Schonfeld 2011). Like the U.S., the UK has provided tax breaks to individuals investing in small businesses in an attempt to get capital flowing again following the credit crisis. The Enterprise Investment Scheme will cut the tax burden on seed stage investments by 50 % and will eliminate capital gains taxes completely on some types of business investments for a short time period (HM Revenue and Customs).

6.6.3 Consistent and Impartial Rule of Law

Investors also depend on stable and predictable laws and regulation and their fair implementation. Regulations must change from time to time to keep up with a changing environment and make use of best practices that have proven their effectiveness in other countries; however, dramatic changes impose costs on businesses and also increase uncertainty about how to comply with a new requirement and how it will be enforced.

The law must also be applied evenly and predictably. Laws that are rarely enforced provide opportunity for selective enforcement. When a law is rarely enforced, it encourages non-compliance and weakens the rule of law. As businesses slacken their compliance with the law, it leaves them vulnerable to unpredictable judgments from bureaucrats, who might have other motives for enforcing a law (such as extracting a bribe, or assisting "favorite" parties). This also leaves businesses guessing about which laws they must comply with.

The ability to obtain licenses is also important for a business and can relate to many areas of work such as construction, handling chemicals or substances, and using certain types of equipment. Delays in obtaining a license can hurt a business through slowing work, idling resources, and ultimately increasing the time it takes for a product to reach the market. Demands for illegal facilitation fees in order to obtain a license more quickly pose unpredictable costs for businesses and damage trust in a government's ability to govern fairly. Any unpredictability in regulation and enforcement increases risks to businesses and creates an environment that is unattractive to investors.

An effective and impartial court system is also necessary for a healthy business environment. The ability to recoup an investment is a primary concern of any investor. Whether it is an investor trying to enforce a contract with a company or a bank attempting to obtain loan repayments, an effective court system is needed to protect an investor. Delays in adjudication draw out legal disputes and increase uncertainty for businesses. Foreign investors also need to be sure that in a legal dispute, a court will treat a foreigner and a local national equally (legal impartiality). If a court system is seen as slow or corrupt, an investor would view this as a potential liability.

Capital is becoming increasingly mobile, and investors will attempt to seek out an investment that yields the best return for the least amount of risk, regardless of whether the opportunity is at home or abroad. Capital flight is often seen as a vote of no confidence in investment opportunities in a country's economy, while strong capital inflows generally indicate investor optimism. A country must reduce risks posed by corruption, a weak rule of law, informality, and poor public institutions in order keep capital from fleeing its borders and to attract new capital from abroad. Considering the very significant risks of funding an innovative, knowledge-based business, any added risks regarding the business climate and the environmental conditions discussed above could quash the potential for high-risk investment.

6.7   Approaches to Supporting High-Risk Finance

Given the increased risk (or even uncertainty) that investors may have toward taking a stake in a company in an emerging market, a government may need to intervene by subsidizing financing or absorbing some of the risk of the investment. These sorts of programs are commonly found in developed markets in North America and Europe since these governments have long recognized the valuable role of small businesses and startup companies in creating job growth and have invested public funds to support their development. The earliest instance of such public support may be the American Research and Development Corporation set up in Massachusetts under the heavy tutelage of the State government in 1946.

Yozma: Israel's Big Bet

Over the past 20 years, Israel has experienced one of the most dramatic transformations from a financially constrained economy with low levels of innovation to a dynamic and highly innovative economy with the highest R&D expenditures as a proportion of GDP in the OECD (OECD 2011). In order to achieve this, the government undertook a bold economic liberalization scheme to shift more capital into the hands of private investors and underwrote the venture investments of experienced foreign venture capital funds in Israeli startups so that domestic investors could learn from the best. In 1993, the Israeli government invested $ 100 million into a program called Yozma ("initiative" in Hebrew) that matched investments of foreign venture capital funds in Israeli businesses. Yozma pre-negotiated buy out prices for the government's shares in an investment so that the venture capital fund could buy out the government if the investment was successful (Gilder 2009). This sweetened the deal for foreign investors since they would be able to assume the profits from the government's shares if the startup started turning a profit without bearing the risks of losses from those shares if the investment failed.

This was a bold and potentially risky strategy since it shifted the rewards for success to the private investors and placed a greater the burden for losses on the government and taxpayers. But it also brought experienced foreign investors into the Israeli market to make investment decisions on Israeli firms and teach local investors how the best VC funds operate. After the success of the first $ 100 million in drawing foreign VC investors, the program grew to $ 210 million to take advantage of the international interest (High Tech Industry Association). These measures helped form local venture capital funds, but it was not until the early 2000s that reforms channeled enough funds into these companies that risk financing in Israel really took off. A series of privatizations and reforms to state pension funds shifted capital away from state bureaucracies and into the hands of private citizens and investors.

The venture capital funds created in the 1990s through Yozma became an avenue for investment for these new streams of capital (Gilder 2009). The Israeli government's bet on bringing foreign venture capital firms into the country has paid of tremendously: from 1991 to 2000, venture capital investment grew from $ 58 million to $ 3.3 billion (Gilder 2009). From 1997 to 2007, Israel's share of GDP devoted to research and development jumped from 2.97 % to 4.76 %. This number has since declined a bit since the global economic crisis, but it still remained the highest in the OECD in 2011 at 4.25 % (OECD 2011). Israel's venture capital spending per capita is the highest in the world at $ 142, around twice the amount in the United States (Vilpponen 2011).

Publicly Funded Venture Capital in the United States

Several U.S. States have devoted public money to venture capital funds to help promising new businesses grow and create jobs for their residents. One of the first examples of these state funds is the Massachusetts Technology Development Corporation (MTDC), established in 1978 by the state legislature to invest in new technology-based enterprises. An independent board comprised of experienced venture capitalists manages the fund and uses money from the legislature to invest in firms seeking $ 2–3 million in funding. By targeting funds in this range, the MTDC plays a crucial role in bridging the funding gap that exists between angel investors and venture capital funds. Over the lifetime of the fund, the MTDC has invested around $ 83 million into 133 companies that currently employ 7,500 people and maintain yearly payrolls that amount to $ 612 million (Kirsner 2011). In addition to creating jobs, the fund has been a good business decision for the state: it has an average internal rate of return of 16.5 % that allows the MTDC to reinvest in new firms without added state support (MTDC 2011).

The State of Connecticut formed its venture capital fund, Connecticut Innovations, in 1989. Through its return on investments, Connecticut Innovations has operated without further financial infusions from the state since1995. The fund operates as an independent corporation with a board that is appointed by the governor and the legislature. Connecticut Innovations' investments range from $ 15,000 to $ 1 million, providing funding to a wide variety of businesses in different funding stages. The fund has leveraged private sector investments in excess of $ 1 billion and has led to the creation of 5,000 jobs in the state.

The State of Maryland created the Maryland Venture Fund in 1993 to target early stage seed companies. The fund has brought in 62.5 million in revenue over the life of the life of the fund while only costing the state $ 41 million to run. Since 1993, 23 of its investments have either gone public with IPOs or have been acquired by a larger firm (Maryland 2011).

There are also many other states that have publicly funded venture funds which vary substantially in terms of size and focus. Apart from direct investments, States also use a variety of other means to encourage private investment in early stage companies such as providing economic incentives, assisting the development of angel networks, and providing investor education. Additional information on these programs can be obtained from the National Governors' Association.

6.7.1   Research and Development Subsidies, Microfinance, and Small Business Support

Germany, both through its federal and länder governments, supports innovative companies through subsidies for research and development. While these subsidies do not specifically target high-risk financing, by subsidizing some of a company's costs, the lower operating costs should make a company less risky and should make any private investment in a company go further. An analysis of the effectiveness of the R&D subsidy found that it did increase innovation in the companies that took part in the program in general (Schneider and Veugelers 2010), but it did not increase innovation at Young Innovative Companies (an EU definition for startups that spend more than 15 % on R&D and are less than 6 years old). An increase of innovation on average for companies receiving the subsidy might be sufficient reason to continue the program, but the authors suggest that a more targeted program might better address the specific needs of a startup.

Evaluations of business support programs in Italy provide support for the notion that targeted support is more effective. While Italy's programs do not focus specifically on innovative or new technology based firms, they do provide an opportunity to evaluate the differences between automatic and selective support. In a study of several of Italy's business support policies, those programs that used a selective review process to award funding resulted in greater business growth and job creation than those programs that distributed funds through an automatic qualification process (Colombo et al. 2008). Further, younger firms grew and created more jobs than established firms. The study found that selective funding was most effective when it was paired with firms that were both young and innovative.

There are a couple of reasons why a selection process might yield better results. While the statistical model used in the study attempts to compensate for selection bias (i.e. those that are more qualified are more likely to be selected for funding, and because they are more qualified, they are also more likely to succeed as businesses), it may still be a factor in the firms' success. When a firm wins funding, it also may act as a mark of approval for the business. Because government program managers have evaluated the company on a number of criteria, this may signal to investors that a selected company is a worthwhile investment. In this way, the selection process is acting to mitigate the information asymmetries that often discourage investment in new innovative businesses.

Conversely, microfinance, which attempts to encourage entrepreneurship on a very broad scale, is not likely to be effective at supporting innovative companies. Microfinance institutions require a large customer base because the value of an individual loan is so small. They do not specifically target innovative firms, nor do they necessarily target firms at all. In many cases microfinancing is often used for smoothing cycles in consumer spending in poor populations, such as using the financing as a stopgap measure when funds run out. Though microfinancing may be of substantial value in helping provide resources to people in times of hardship, it does not tend to encourage entrepreneurship or more small businesses (Dichter 2007). Because it has a minimal impact on small business formation and growth, its effect on highly innovative small businesses, a small subset of businesses in general, would be even further diluted. Additionally, administration costs for microfinance have to be kept to a minimum because the small values and the high volumes of the loans require a highly efficient loan distribution system in order to keep administrative costs from eating away at profits, or in most cases for microfinance loans, to minimize losses. This prevents a microfinance institution from providing the type of business training that is considered highly effective in fostering well-functioning small businesses. While microfinance may have a place in development, it is not an effective tool for providing finance to innovative businesses.

A successful example of a research subsidy mechanism is the Small Business Innovation Research (SBIR) program in the US.[3] The program requires that agencies that spend over $ 100 million annually on R&D to devote 2.5 % of their budget to SMEs with fewer than 500 employees. The program was initially conceived with a three stage funding process. The first stage provides up to $ 1,500,00 for 6 months to study the feasibility and potential of an R&D project. The second stage provides up to 1,000,000 for a year to continue the R&D efforts based on the results of

---

[3] SBIR: http://www.sbir.gov/

stage I. Stage III does not provide any funding, instead the agencies or commercial entities that will benefit from the R&D are expected to pay for the research from non-SBIR funds if they see merit in the research (SBIR 2011). A critical reform to the program in 1995 allowed for an expedited review process for stage II if a company was able to obtain outside matching funds. They also receive bridge funding of $ 30,000 to $ 50,000 between stages I and II to prevent any funding gaps in the interim. This resulted in the program attracting younger companies that are in the most need of stable capital.

A study revealed that the SBIR program resulted in public benefits that would not have resulted from private funds had support from government grants not been present. It found overall that the program was socially valuable and a good use of public funds (Link and Scott 2000). While developing countries may not be able to take the same approach due to small research budgets, the important feature in improving the performance of this program was minimizing gaps in funding. For funding projects that are taking place in stages, minimizing gaps is an important part of program design.

6.7.2 Case: Public Support Programs for High-Risk Financing in Finland

Finland has developed a publicly-funded system of institutions designed to support new startups and help carry them over the valley of death. Though these institutions overlap in several areas, each institution plays a role in supporting young companies at a particular funding gap. The system provides the potential for a startup to rely on government support or government-facilitated support starting at pre-seed business planning and continuing all the way until the company has moved beyond early stage development to a size that is more attractive to private venture capital.

This approach involves a strong role for the government, but seems to work well in Finland, which has a high number of high tech innovative companies and spends the second highest amount of its GDP on R&D in the OECD. Finland has also achieved one of the highest venture capital expenditures in Europe at roughly $ 60 per capita. For comparison, the U.S. spends $ 67 per capita (Wagner and Laib 2011). Estimates for venture capital expenditures across the entire EU reach as low as $ 7 per capita (Vilpponen 2011) (Table 6.3).

Table 6.3 Finland's High Risk Finance Institutions

Sitra	Tekes	Finvera	Finnish industry investments
Pre-seed	Seed	Early/startup	Startup
Grants and loans for creating business plans and evaluating	Loans for startup costs and grants for research and development	Direct equity investments of up to 500 K €	Encouraging private equity involvement through investing in private venture funds and co-investing in private venture

Sitra, the Finnish National Fund for Research and Development, is a fund directed by the Parliament that engages in a wide range of activities from investing in venture funds to financing individual businesses. Sitra in recent years has focused its efforts on the earliest pre-seed stages of new business. Sitra, in partnership with Tekes, provides hybrid grant-loan support up to 40,000 € to entrepreneurs. This funding helps an entrepreneur purchase consulting services to create a business plan and evaluate the feasibility of a company's R&D goals (Maula and Jääskeläinen 2007).

Tekes, The Finnish Funding Agency for Technology and Innovation, is primarily geared toward supporting new knowledge-based enterprises through research and development grants and loans, but also provides up to 80% of the startup capital, up to 100,000 €, for new businesses through unsecured loans. This type of loan can help overcome one of the most difficult points in the valley of death through providing necessary funds at the earliest stages of a company when private capital is scarcest. Further, Tekes provides technical assistance to potential applicants and provides loan support for developing business models (Maula and Jääskeläinen 2007).

Finnvera is Finland's Export Credit Agency, which, in addition to helping secure Finnish companies against the risks of internationalizing and expanding exports, engages in venture capital investments and the support of SMEs. Finnvera supports innovative knowledge-based startups through a subsidiary called Avera which makes direct investments in companies without private partners. Avera targets its funding to bridge the gap between R&D funding and venture capital funding. Avera funding frequently follows Tekes funding to sustain businesses that are attempting to commercialize a product and attain private sector venture funding. The size of the investment, up to 500 K €, is large enough to act as bridge funding to move a company beyond the seed stage (Finnvera).

Finnish Industry Investment Ltd. started out primarily as an indirect investor in new companies through making equity stakes in venture funds. Its purpose was to partner with private equity and encourage the creation of new venture capital funds in order to boost the involvement of private capital in early stage financing. After encountering difficulty in convincing private investors to co-invest with a government controlled entity in venture capital funds, FII began investing directing with private investors in new companies on a matching basis. When making co-investments with private investors, FII typically lets the private sector partner lead on the investment and does not accept a board seat in the company. About half of FII's funds are invested in venture funds and half are invested directly in companies (Finnish Industry Investment).

Finland's public support of young innovative startups features significant government support at the early stages, but then shifts government involvement to encouraging private capital formation and investment once a company moves beyond the seed stage. This focuses government support where market failures are the greatest, but it also relies on government officials picking winners and losers among applicants for support. The success of this model depends on the experience and capabilities of the government institutions implementing these programs.

6.8   Recommendations

In presenting recommendations for promoting high-risk financing, it is important that we not put the cart before the horse. Angel investors and venture capital funds are limited primarily by a lack of what investors "perceive to be promising entrepreneurs and high-potential firms suitable for investment" (OECD 2004). "[T]he most fundamental requirement for facilitating funding for innovative SMEs is to create an economic and institutional environment that is conducive to entrepreneurship and innovation" (OECD 2004). Developing opportunities for investment must precede any effort to stimulate high-risk financing.

The recommendations below apply to moderately-developed countries seeking to encourage high-risk finance in domestic markets. Early-stage investors need physical proximity to monitor their investments, making foreign investment in seed companies and early-stage startups less likely. Foreign investment aimed at expanding a later-stage company that already has revenue streams is a likelier possibility. It might also make a foreign investor more confident about investing in a developing market since the firm has already proven that it can flourish against the added challenges in such a market. Investors need less encouragement to invest in successful companies, so these recommendations will focus on developing small businesses that have not yet proven themselves.

- *Support companies at the seed stage.* The riskiest stage of investment is at the very early pre-seed stage when an idea for a company is just taking shape. The seed stage suffers from a dearth of investment and represents an opportunity for government support to help deserving companies bridge the gap. Government money should be used to leverage private funds so that the impact of public investment is maximized. Potential methods for doing this are direct grants, matching requirements, loan guarantees, subsidies, and tax credits.
- *Mitigate the costs of failure.* By making it possible to recover from failure, entrepreneurs can learn from their mistakes and try again. Government programs that subsidize funding for startups should develop appropriate penalties for failure that balance make failure costly enough to provide disincentives for reckless borrowing with the need to make sure that an entrepreneur can recover from a failed business.
- *Make sure that everyone plays by the same rules.* Government policies should attempt to create a level playing field so that businesses can prosper and fail not on the basis of favoritism, their connections, or their willingness to pay bribes, but on their ability to compete. True competition can only take place when everyone plays by the same rules. Enforce the rule of law consistently and equitably, and vigilantly protect intellectual property rights.
- *Provide training and require monitoring and management assistance.* Capital alone will not make a business successful. Part of the value of angel investors and venture capital funds is the management assistance and monitoring that they provide. Combining debt-based financing with the management and oversight typically found in equity-based financing will make loans a more productive method for financing startups.

- *Be selective and use meritocratic criteria in choosing companies to fund.* Programs that distribute funds based on a review-based meritocratic selection process are more successful than those that provide funds based on criteria that confer automatic eligibility. This helps ensure the efficient use of public funds by directing funds toward only the most promising firms and by limiting bankruptcy losses. The vetting process before funds are awarded helps decrease the information gap and risk that other investors face when they are deciding whether to invest.
- *Systematize seed and venture capital financing.* Entrepreneurs need to know what to expect when they start a business and how they are going to obtain funding at each stage along the way. Systematize financing by organizing institutions and private investors so that the methods for financing are clear at each stage of development. This will make financing easier and more practicable for both entrepreneurs and investors alike and avoid ad hoc arrangements that can make transactions more cumbersome and unpredictable.

References

Ben-Ari, G., & Vonortas, N. (2007). Risk financing for knowledge-based enterprises: mechanisms and policy options. *Science and Public Policy, 34*(7), 475–488.

Black, B., & Gilson, R. (1998). Venture capital and the structure of capital markets: Banks versus stock markets. *Journal of Financial Economics, 47,* 243–277.

State of Connecticut, Connecticut Innovations. (2011). About Connecticut Innovations. http://www.ctinnovations.com/AboutUs.aspx. Accessed 9 Dec 2011.

Colombo, M., Giannangeli, S., & Grilli, L. (2008). A longitudinal analysis of public financing and the growth of New Technology-based Firms: Do firms' age and applicants' evaluation methods matter?. Unpublished Manuscript.

Das, S., Jagannathan, M., & Sarin, A. (2003). Private equity returns: An empirical examination of the exit of venture-backed companies. *Journal of Investment Management, 1*(1)

Dichter, T. (2007). A second look at microfinance: The sequence of growth and credit in economic history. Development Policy Briefing Paper No. 1. CATO.

European Commission. (2002). Enterprise Directorate-General, "Benchmarking Business Angels. http://ec.europa.eu/enterprise/newsroom/cf/_getdocument.cfm?doc_id=1120. Accessed 8 March 2012.

Finnish Industry Investment. About Us, Finnvera, "Operating M. http://www.finnvera.fi/eng/Venture-Capital-Investments/Operating-model. Accessed 8 March 2012.

Gilder, G. (2009). Silicon Israel: How Market Capitalism Saved the Jewish State. *City Journal, 19*(3). http://www.city-journal.org/2009/19_3_jewish-capitalism.html. Accessed 8 March 2012.

Gompers, P. (1995). Optimal investment, monitoring, and the staging of venture capital. *Journal of Finance, 50,* 1461–1489.

Gompers, P., & Lerner, J. (1999). An analysis of compensation in the U.S. venturecapital partnership. *Journal of Financial Economics, 51,* 3–44.

Hall, B., & Lerner, J. (2009). *The financing of R & D and innovation. Handbook of the economics of innovation.* North Holland: Elsevier.

Hellmann, T., & Puri, M. (2000). The interaction between product market and financing strategy: The role of venture capital. *Review of Financial Studies, 13,* 959–984.

Her Majesty's Revenue and Customs. Enterprise Investment Scheme (EIS). http://www.hmrc.gov.uk/eis/. Accessed 8 March 2012.

Kane, T. (2010). The importance of startups in job creation and job destruction. The Kauffman Foundation. http://www.kauffman.org/uploadedFiles/firm_formation_importance_of_startups.pdf. Accessed 8 Dec 2010.

Kirsner, S. (2011). What will the future of the Mass. Technology Development Corp. look like? Boston Globe, July 27, 2011.

Kortum, S., & Lerner, J. (2000). Assessing the contribution of venture capital to innovation. *RAND Journal of Economics, 31,* 674–692.

Canada, Industry Canada. (2001).Gaps in SME financing: An analytical framework. Ottawa, Canada.

Landier, A. (2006). Entrepreneurship and the Stigma of Failure. New York University. http://pages.stern.nyu.edu/~alandier/pdfs/stigma9.pdf. Accessed 22 Aug 2011.

Link, A., & Scott, J. (2000) Estimates of the social returns to small business innovation research projects. In C Wessner (ed.) *The small business and innovation research program: An assessment of the department of defense fast track initiative.* National Academy Press.

State of Maryland, Maryland Department of Business and Economic Development (2011). Annual Fiscal Status Report Fiscal Year 2011. http://www.choosemaryland.org/aboutdbed/Documents/ProgramReports/2011/IFGAnnualReportFY11.pdf. Accessed 9 Dec 2011.

Mani, S., & Bartzokas, A. (2002). *Institutional support for investment in new technologies: The role of venture capital institutions in developing countries. institute for new technologies discussion paper no. 2002–2004.* Maastricht: The United Nations University.

Maula, M., Gordon, M., & Jääskeläinen, M. (2007) Public financing of young innovative companies in Finland. Government of Finland, Ministry of Trade and Industry. http://ktm.elinar.fi/ktm_jur/ktmjur.nsf/All/7272F51ED5707869C2257284001F5863/$file/jul3elo_2007_eng_netti.pdf. Accessed 8 March 2012.

Megginson, W., & Weiss, K. (1991). Venture capital certification in initial public offerings. *Journal of Finance, 46,* 879–893.

National Venture Capital Association. (2011). VC Investments Q2 '11–Money Tree—National Data. http://www.nvca.org/index.php?option=com_content&view=article&id=79&Itemid=103. Accessed 22 Aug. 2011.

OECD, OECD Factbook (2011–2012). December 11, 2012. http://www.oecd ilibrary.org/docserver/download/fulltext/3011041ec068.pdf?expires=1331235464&id=id&accname=freeContent&checksum=92858EBF8193E3248193900F70C2970A. Accessed 8 March 2012.

OECD, Organization for Economic Cooperation and Development, (2004). Promoting Entrepreneurship and Innovative SMEs in a Global Economy: Towards a More Inclusive and Responsible Globalization.

OECD (2007). Improving the entrepreneurial finance: Venture Capital and access to credit for SMEs prepared by AlissaKoldertsova. http://www.oecd.org/dataoecd/48/55/38515315.pdf. Accessed 9 Dec 2011.

PWC/NVCA (2014).PricewaterhouseCoopers National Venture Capital Association MoneyTree Q4 2013/ Full-year 2013 Data provided by Thomson Reuters. https://www.pwcmoneytree.com/MTPublic/ns/moneytree/filesource/moneytree/filesource/exhibits/Q42013NatlAggSpreadsheet.xlsx. Accessed 9 Feb 2014.

Schneider, C., & Veugelers, R. (2010). On young innovative companies: Why they matter and how (not) to policy support them. *Industrial and Corporate Change, 19*(4), 969–1007.

Schonfeld, E. (2011). Attention, Angel Investors: You Have Until Jan. 1 To Lock In 100 % Tax-Free Capital Gains On Startup Stock, TechCrunch. http://techcrunch.com/2011/12/13/angel-investors-tax-free-capital-gains-startup-stock/. Accessed 8 March 2012.

Sohl, J. (2003). The Angel Investor Market in 2003: Investor Activity and Growth Prospects, Center for Venture Research.

Sohl, J. (2010). The Angel Investor Market in 2009: Holding Steady but Changes in Seed and Startup Investments, Center for Venture Research.

Sohl, J. (2011). The Angel Investor Market in 2010: A Market on the Rebound, Center for Venture Research.

United States, Small Business Administration. (2011). The SBIR program. http://www.sbir.gov/about/about-sbir. Accessed 25 Oct 2011.

Vilpponen, A. (2011) VC Per Capita: Europe $ 7, US $ 72, Israel $ 142, ArcticStartup, June 15, 2011. http://www.arcticstartup.com/2011/06/15/vc-per-capita-europe-7-us-72-israel-142. Accessed 8 March 2012.

Wagner, S., & Laib, L. (2011).Theory vs. reality: Venture capital in Europe, Investiere, Verve Capital Partners AG. http://www.startupticker.ch/getattachment/News/January-2012/Switzer-land-with-the-highest-Venture-Capital-spend/VCP-Study-Venture-Capital-in-Europe-2011. pdf.aspx. Accessed 8 March 2012.

World Bank. (2011). Doing Business. http://www.doingbusiness.org/data/exploreeconomies/ lebanon/#getting-credit. Accessed 9 Dec 2011.

Chapter 7
Intellectual Property, Standards

Jeffrey Williams and Anwar Aridi

7.1 Introduction

This chapter examines intellectual property and standards, two important elements of the innovation landscape. First to be highlighted will be intellectual property (IP), a set of rules and institutions designed to foster innovation and ideas. We will discuss different forms of intellectual property protection such as patents, copyright, trademarks, and trade secrets. The chapter will then discuss standards and their role in domestic innovation and in international trade. Standards have the potential to boost innovation, but also have the potential to stifle domestic industrial creativity.

7.2 Forms of Intellectual Property Protection[1]

Intellectual property (IP) is an idea, or a collection of ideas, produced in the expectation of direct or indirect economic gain. Intellectual property regimes are nation-level mechanisms designed to protect these ideas by assigning control over their use to their creator. Generally, governments are concerned about ideas in so far as they

---

Adapted from a chapter of the *Innovation Policy Handbook* report composed for the World Bank (2012). Original unpublished and available upon request.

---

[1] This section relies extensively on Scotchmer (2005).

J. Williams (✉)
Center for International Science and Technology Policy, The George Washington University, Washington, D.C., USA
e-mail: williamsjeffl@gmail.com

A. Aridi
SRI International & Trachtenberg School of Public Policy and Public Administration, The George Washington University, Washington, D.C., USA
e-mail: aridi.anwar@gmail.com

© The Editor(s) 2015
N. S. Vonortas et al. (eds.), *Innovation Policy,* SpringerBriefs in Entrepreneurship and Innovation, DOI 10.1007/978-1-4939-2233-8_7

are used to spur innovation and economic growth, and thus the implementation of IP regimes to protect those ideas will have a strong bias towards fostering economic growth.

There are four methods of formal intellectual property right (IPR) protection: patents, copyrights, trademarks, and trade secrets.[2] Patents and copyrights will form the bulk of the IP discussion in this chapter as they are the most complex forms of IP protection related to innovation and trade issues.

- *Patents* are offered to stimulate production of new ideas. They work by providing a limited-time right of exclusion to the creator of an idea. Violators of this right of exclusion often must pay a fine or other penalty to the owner of the idea. Patents must be applied for and must prove (a) patentable subject matter, (b) utility, (c) novelty, and (d) non-obviousness. Patents last twenty years from the date of filing.
- *Copyrights* are offered to stimulate expression. Protection is automatically given to any original work of authorship such as books, software, music, and movies. Copyright gives holders the right to copy, reproduce, distribute, adapt, perform, or display their works. Importantly, copyright allows creators of ideas to prevent others from selling reproductions of the original idea but does not prevent others from expressing similar yet distinct ideas. Copyright lasts for the life of the life of the author plus 70 years.[3]
- *Trademarks* refer to a distinctive word, phrase, symbol or design, or a combination of these that identifies and distinguishes the source of the goods of one party from those of another. Trademarks allow markets to function smoothly by supplying information to buyers. Trademark registration enables the registrant to exclude others from using the trademark in ways that could cause confusion in the marketplace.
- *Trade secrets* protect various types of firm-specific technical and business knowledge. They refer to confidential business information that provides an enterprise a competitive advantage. Trade secrets may include manufacturing or industrial secrets and commercial secrets such as production technologies, sales methods, distribution methods, consumer profiles, advertising strategies, and lists of suppliers or clients. Trade secrets are not registered, but are protected through security procedures and confidentiality agreements. Proof of violation requires evidence that information was obtained by improper means (e.g. industrial espionage). Trade secrets that are discovered accidentally (e.g. in the process of reverse engineering) are not protected.

Informal IPR protection includes the use of secrecy, complex routines, and speed in technological advancement.

- *Secrecy* refers to the ability to keep abstract or applied technical ideas secret. It is probably the most effective informal method for retaining intellectual property proprietary. Whether knowledge can be kept secret is a matter of the technology

---

[2] The legal details here are based on the U.S. model, but similar criteria apply in other nations.

[3] The importance of copyrights has increased tremendously in the era of the knowledge economy. In fact, various experts argue that the extensive attention on patents in the detriment of other forms of IP protection may be increasingly misguided (Wunch-Vincent 2013).

involved. If reverse engineering is relatively easy, then formal protection is necessary. Secrecy helps avoid the revelation of information to prospective competitors through published patents, thus, avoiding "inventing around" patents.

- *Complex routines* refer to the situation where competitive advantage consists of accumulated experience (routines). In such cases firms may consider themselves reasonably protected from imitators. This may be the case of established companies that deal with complex, systemic products.
- *Speed* refers to the practice of speedy development of ideas in rapidly changing technologies. By the time a competitor obtains the ability to copy an existing product or process, the IP owner has developed a more advanced product or process that diminishes the market importance of the first product or process.

Formal IP protection is based on legal measures. With the exception of trade secrets, all other forms of formal protection require the provision of detailed information about the object of protection. Inventors will rely upon them as long as the explicit and implicit costs of doing so appear justified by the potential benefits. Frequently they decide not to use formal protection in order to avoid revealing too much information. Informal protection thus seems to be used extensively.

7.3   Intellectual Property in the Innovation Ecosystem

Nations have separate sets of regulations to protect IP and physical property because they are fundamentally different types of valuable goods. Today's global economy is built largely off of the exchange of knowledge-based goods (Harris 2001; Powell and Snellman 2004; Godin 2005; OECD 2013). These are goods with significant value added from scientific research and the application of learning and technical ideas, and it is the knowledge behind these goods that is protected by IP regimes.

But how exactly are IP and physical property different? For one, knowledge is a cumulative good, that is to say that knowledge follows a path dependency (Malerba et al. 1999). A country may aspire to be world leader in nanotechnology, but if there is no prior accumulation of nanotechnology expertise in that country, then it will be difficult to innovate in that arena. In contrast, a country with a tradition of agricultural excellence likely is able to swiftly and efficiently adopt and adapt new agricultural techniques. A second characteristic of knowledge is that it is irreversibly transferrable; once someone learns something, that knowledge cannot be taken away. From an industrial standpoint, this means that personnel do not forget what they know when they move to a different company or country, and once a competitor understands a firm's internal processes, that understanding cannot suddenly disappear. Third, knowledge is subject to increasing returns to scale, meaning that outputs stemming from knowledge increase at a proportionally greater rate than an increase in inputs for its production. It is this characteristic of increasing returns that allows knowledge-based economies to be so dynamic.[4] Fourth, knowledge has high initial costs and much lower marginal costs of production (in some cases close to

---

[4] It is this feature of knowledge that underlies the explanations of economic advance of new growth theory discussed in Chap. 2 of this volume.

zero). Piracy of software receives so much attention exactly for this reason; it takes significant resources to create, and virtually no resources, training, or skills to illegally copy and resell. The same holds for music and many other knowledge-based goods.[5]

The major impact of IP, specifically patents, is in the ability of the owner of a piece of knowledge to appropriate rents stemming from the development and commercialization of that knowledge. In order for company A to invest in the creation of a new piece of knowledge (an invention), that company must have reasonable expectations that it can profit from the invention, and thus it wants to be sure that it can appropriate that invention and keep it private. Employees can jump to new firms and carry knowledge over with them and competitors can reverse engineer a finished good. Secondary parties then could exploit the knowledge company A paid to create.[6] As one of the characteristics of knowledge is high up-front costs for its production and low marginal costs for its reproduction, competing producers of products largely based on this specific piece of knowledge appropriated without proper payment to the owner are now at a huge cost advantage which can allow them to profitably undercut company A in the market.

It is critical that any IP regime be reflective of and incorporated into the overall innovative framework of a nation; it is not a stand-alone mechanism (World Bank 2010). For instance, there could be strong IP laws in a country, but if the legal system was not able to enforce those laws due to lack of resources, training, or enforcement authority, then it would be as if the laws did not exist; there would be no encouragement for innovation.

7.4 Intellectual Property and Development

The previous section asked what the role of IP in the innovation framework was. Now we ask a more nuanced question: what is the role of IP in the innovation frameworks of countries at varying stages of development? What does it mean to the global economy that in 2012, for the first time, China, not the US, held the top positions for both destination and source of patent filing? How do we interpret the recent spike in patent application filing abroad of BRICS origin?

Let's start from a basic idea, at the core of economic thinking since at least Arrow (1962). The economic rationale for IP protection rests on the trade-off between allocative efficiency and dynamic efficiency. Simply put, allocative efficiency means that, assuming no future inventions, the efficiency of the economic system is maximized by spreading knowledge around: everybody knows everything. Dynamic efficiency changes the basic assumption: if there is knowledge to be created and things to be invented in the future, then some sort of monopoly power expectation must be created to incentivize the necessary expenditure from individuals or organizations. In the extreme, allocative efficiency corresponds to the absence of IP protection. In

---

[5] For more details on knowledge as an economic input see Romer (1996), OECD (1996), Grand-strand (1999).

[6] Albeit at a cost. Research has shown that the costs of imitation vary across industries and across activities and can be significant. See Mansfield (1985), Mansfield et al. (1981), Levin et al. (1987).

contrast, dynamic efficiency requires such protection. The problem is that in actuality we need both: prospective inventors must have some guarantee of legal appropriability, whereas the economic system will progress with people other than the inventor eventually getting hold of the specific knowledge. Extant IPR regimes have sought the middle ground by providing monopoly rights for new patentable ideas but for a price and for a limited time period after which the knowledge becomes public.[7,8]

China Assumes Leadership

For the first time, China holds the top positions for both destination and source of patent filings. In 2012, for the first time, residents of China accounted for the largest number of patents filed throughout the world. In addition, State Intellectual Property Office (SIPO) of the People's Republic of China accounted for the largest number of applications received by any single IP office. Residents of China filed 560,681 patent applications; this compared with those filed by residents of Japan (486,070) and residents of the United States of America (US, 460,276). Similarly, SIPO received 652,777 applications, compared to 542,815 for the USPTO and 342,796 for the Japan Patent Office (JPO).

Source: World Intellectual Property Organization (WIPO) 2013 World Intellectual Property Indicators (Fig. 7.1)

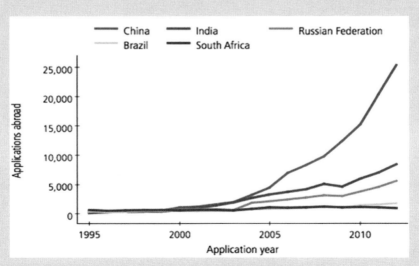

Fig. 7.1 Patents of BRICS Origin. (Source: WIPO Statistics 2013, p. 60)

---

[7] For an excellent historical exposition of how IPRs came to be and what they mean see David (1992).

[8] Other more esoteric issues are also relevant here and have been widely discussed by economists regarding the warranted strength of the patent system including the breadth of protection (how broad a patent is) and the number of claims on a single patent. We refrain from these topics herein.

In addition, one might ponder the question whether all countries need the same extent of intellectual property protection. In particular, could it be that countries at different stages of economic development would benefit from different degrees of protection? The answer may well be affirmative given that countries at the top of the development ladder base their competitiveness on the creation and application of state-of-the-art technologies based on advanced scientific research, whereas those on the other end often find it more profitable to concentrate limited resources to the identification, adaptation and adoption of extant technologies and broad dissemination of information (Lall 2003). In other words, advanced nations are expected to push for stricter IPR regimes focusing primarily on dynamic efficiency whereas lesser developed nations are expected to push for laxer IPR regimes as they are primarily concerned with allocative efficiency.

The question above proves, in fact, to be one with no easy answer. Intellectual property has different impacts on innovation for countries at varying levels of development largely because of trade. In less developed countries innovation occurs primarily through importation of technology from more developed nations (World Bank 2010). Firms engaged in selling technology in international markets want to be assured of protection for their investments, and most are more eager to sell in countries where a reasonable IPR regime is in place (Branstetter et al. 2005). The presence of a functioning IPR system is a strong market signal to prospective market entrants.

Are, then, countries that lack a reasonably functioning IPR regime effectively cut off from technology imports? Not at all, as firms will export technology in a format appropriate to a customer nation's level of absorptive capacity. Two characteristics determine absorptive capacity. The first is appropriability, which conditions technology transfer on the ability of domestic R&D concerns to incorporate foreign technology and learning into their own production processes. The second is usability, which argues that the level of technology imported depends on the level of development of the target country (Gibson and Smilor 1991; Javocik 2005; Park and Lippolt 2008). For example, a less developed nation may import semi-conductor technology in the form of finished computers (usability), but there might not be any domestic firms that could import the latest semi-conductor know-how and use it to develop a new computer themselves (appropriability). Firms, then, generally will export finished, high tech goods to areas with weak IP protection and are more comfortable exporting know-how in a country with stronger IP protection.[9] Know-how exports may be in the form of a factory or processing facility, or a collaborative venture with local firms, or outright licensing agreement.

Just as firms have determinants for the type of technology they are willing to export, developing countries have determinants for the type of technology that is imported. Less developed nations will see much more efficient outcomes by focusing on importing technology rather than creating the infrastructure to create it locally (WIPO 2011). Moreover, IP regimes that are too stringent for a developing

---

[9] Usability and appropriability are also time specific; as a country develops, its technology capacity changes. See the text box on South Korea's technology development.

nation may lead to technology-associated economic rents being directed to foreign firms (Ganslandt et al. 2005; World Bank 2010), thus decreasing the efficiency of a national system. On the other hand, developed nations will see much more efficient outcomes by pushing the boundaries of a technology through constant innovation (Abel et al. 1989). In this situation, strong IP regimes encourage domestic producers to invest in innovative activities by providing a more secure appropriability of rents.

Overall, the complex set of factors determining the transfer of technology from abroad include the country's IPR system, its position in the global value chain, the size of its existing or prospective domestic markets, and strong public policy preference.

7.5 Determining the Need for and Impact of Intellectual Property

Evaluating IP policy involves understanding the innovation ecology of one's country and the impact on the whole system. Broadly speaking, IP policy affects two populations to varying degrees. First, it affects those entities active in invention and innovation, such as firms, universities, and entrepreneurs. If the policy encourages innovation, and if innovation is associated with economic growth, then IP can affect the economic climate of the entire country. Determining if an IP policy is effective, therefore, involves more than just counting the number of patents, or relying on any single measure of impact. For example, Branstetter et al. (2005) point out that stronger IPRs will attract more technology investment from foreign firms, but that measurement alone does not tell us if the new investment is putting domestic firms out of business, leading to a trade imbalance, or over-burdening existing infrastructure. Second, consumers stand to be affected if stronger IP rules attract more foreign technology imports with the unplanned effect of pricing that technology out of the reach of the domestic consumer (Fink and Maskus 2005).

Developing nations face a clouded path to IP implementation. The pure allocative or dynamic efficiencies discussed earlier will not apply uniformly across their economies. Some areas of technological skill in a developing nation will be far from the cutting edge, while others might be much closer. The rapidly advancing BRIC nations—Brazil, Russia, India, and China—represent this middle ground on a grand scale. All of them, in various fields of technology, are innovative leaders and followers (Tseng 2009). From an IP policy standpoint, this is a difficult position to occupy and all four of these nations have tried varying forms of IP legislation in an effort to encourage simultaneously domestic innovation and foreign technology investment. The act of balancing domestic innovation needs and foreign IP requirements, while stimulating growth at home, have at times attracted sanctions or threats of sanctions from more developed nations (Bird and Cahoy 2007). In general, patent applications have been increasing worldwide with new patent offices emerging as key players such as China's State Intellectual Property Office (SIPO) (Figs. 7.2 and 7.3).

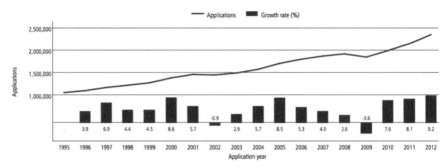

Fig. 7.2 Trend in patent applications worldwide (Source: WIPO Statistics 2013, p. 46) Trend in Patent Application for the Top Five Offices

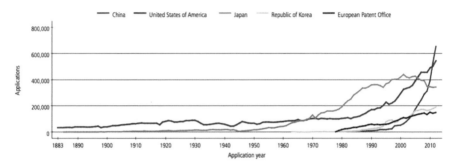

Fig. 7.3 Trend in patent applications for the top five offices (Source: WIPO Statistics 2013, p. 52)

Countries further away from the cutting edge of a technology may find it beneficial, or at least tempting, to relax IP rules and enforcement. While this increases access to knowledge from foreign sources and lowers the barrier to innovation for domestic producers, care must be taken that laws are enforced appropriately. Purposeful lax enforcement of strong rules or creation of weak rules that deliberately allow domestic firms to "legally" appropriate foreign technology can drive away foreign investment and technology and harm the domestic innovation landscape. For example, in the early 1980s, South Korean IP laws tacitly allowed for what essentially amounted to theft of foreign IP. In one instance, trademarks were considered valid only if the brand was familiar to most Koreans, a loophole that meant many foreign-made goods de facto lacked protection. The response of foreign technology suppliers was a round of steep trade sanctions (Ryan 1998).

As the above examples emphasize, designing an IP system that both encourages domestic innovation and supports the legal importation of foreign technology is difficult. It is critical for policymakers in these situations to understand the needs and capabilities of prospective domestic innovators and be able to revisit and rebalance the national IPR regime. The tools to do so are variable. Surveys are a useful tool because they assist in obtaining detailed information such as frequency of patents and copyrights, ease of obtaining IP protection, income generated from

IP-protected goods, whether IP owners consider the process a good investment of resources, manufacturing, marketing, or distribution problems solved or caused by IP, etc. Generally, it is not feasible to survey every entity involved in an innovation ecosystem, but making contact with as many business owners as possible, both innovators and those who use their innovations, is critical.[10]

Another method of determining the effectiveness of an IP policy is through analysis of patent data. Patents have the advantage of being a distinct, quantifiable phenomenon. Patent data can be broken down into a number of useful categories, such as: area of technology, location of inventor, location of owner if different from inventor, and previous knowledge on which the patent is based. The danger of patent data alone, however, is that it lacks context. Mowery and Sampat (2005) describe a good example of the need for context. In the US, there has been a surge in patenting by universities, which many interpret as validating the government's efforts to have universities participate more in the innovation process. But such an increase in patenting may be missing a possible long-term effect in decreased innovation as universities redirect their focus towards short term research.

IP Case—Korean Pharmaceutical Industry and Development

In the 1960s and 1970s, Korea, like many industrialized, developing nations, was building its technology base by copying mature foreign technology. With relatively cheap labor costs, Korean firms were able to produce these mature technologies for domestic and international consumption at competitive prices. As the nation developed economically, however, those labor costs rose. Sensing that this is hardly a long-term strategy, the country relentlessly tried to upgrade. In the 1980s, Korean firms began to manufacture more sophisticated, value-added technological goods with increased technological know-how coming from three sources: copying cutting edge foreign technology, increased spending on R&D, and a base of domestic technology experience developed from copying mature foreign technology.

Intellectual property became a concern for Korea in the 1980s. Prior to that period, the foreign technologies that Korean firms were able to copy were mature, with innovation coming mainly in the marketing and manufacturing processes and costs highly driven by worker wages. IP played a much smaller role in maintaining a competitive advantage then and, thus, foreign firms were less likely to block Korean firms from using that technology. However, once Korean firms began to create and copy more value-added products like pharmaceuticals, they came into more frequent conflict with foreign firms who owned the more advanced IP.

---

[10] Good examples of survey use regarding innovation are the OECD Country Innovation Policy Surveys. For further information see also WIPO (2013) and references therein.

The pharmaceutical industry in Korea grew very rapidly in the 1980s, and this was almost entirely due to the copying of foreign products. At first, Korea officially honored process patents, but not product patents, which allowed domestic firms to jump into high-tech manufacturing once a product's manufacture was deciphered. Meanwhile, Korea's trademark law only allowed trademarks for products that were well-known to the Korean people, thus tacitly allowing the copying of any foreign good. Foreign governments cried foul, and the Korean government created tougher laws. However, enforcement was notoriously lax, and the copying continued. By the end of the 1980s, nearly 90% of the Korean pharmaceutical market was supplied by domestic firms, a percentage much higher than equivalently developed nations at that time. Eventually, threats of sanctions from international partners forced the institution of real IP enforcement.

Today, Korea has a thriving pharmaceutical industry. Certainly, this path of development is not conducive to winning the trust of international partners. It is also important to not allow lax IP enforcement to undercut one's own internal development by de-incentivizing investment in domestic high-tech industries. The Korean experience provides lessons on both the balancing of IP enforcement and technology development as well as the trade problems associated with IP as a nation moves up the development curve.

Sources: Ryan 1998; UNCTAD 2003

The principal take-away is that IP is a necessary but complex policy tool implemented in a complex innovation environment. Not only does the national IPR regime need to be calibrated to encourage domestic innovation and remove barriers to the spread of new technology but, for developing nations especially, it must provide foreign technology providers with confidence that their knowledge investments will be safe. This is a difficult path to navigate, and requires policymakers to pay careful attention to the creation and implementation of laws and institutions. Hence, the efforts of both developed and developing countries to protect their patents in major markets. The figure below shows that the US is still the main destination for filing for top five origins and BRICS origins (Fig. 7.4).

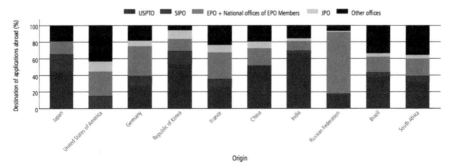

Fig. 7.4 Destination of applications abroad for the top five origins and BRICS origins, 2012 (WIPO Statistics 2013, p. 61)

7.6 Standards

Standards constitute another driver of innovation, and they are becoming more important as markets expand and internationalize. Goods and services are increasingly designed and developed to be sold internationally, thus, required to satisfy a variety of markets and cultures with different requirements on materials and processes and different preconceptions of what is good or bad. "Standards are not only a technical question. They determine the technology that will implement the Information Society, and consequently the way in which industry, users, consumers and administrations will benefit from it" (EC 1996). Standards encompass an increasingly wide range of manufacturing, process, and ethical requirements placed on commercial goods. Standards can be imposed by governments, international bodies, trade associations, or can be the outcome of regular market transactions.

Typically, only those countries on the technological frontier will be able to set the standards for technologies for which they are the primary producers, users, and sellers. Specific technologies develop along a series of steps, one of which involves the intentional or circumstantial setting of standards (Gort and Klepper 1982). Technology followers by definition come upon a technology after it has already been in development for some period of time, and often miss the standard-setting stage. However, specific circumstances occasionally provide the opportunity to developing countries close to the frontier of a specific technological field to set or influence standards. Market size is one of those circumstances. A great example is China's attempt to set its own signal security standard for all wireless devices sold in China, including imports. Makers of wireless devices from other nations balked at this proposal, as it would have created a second set of security standards for makers to meet, upending international markets and virtually forcing foreign firms to provide Chinese manufacturers with proprietary information (Gibson 2007). While the proposal for the new standard was ultimately withdrawn due to international resistance, the chain of events indicated a couple of things: (a) China was advanced enough technologically to set their own wireless security standard; (b) China had a large enough internal market that international makers could not simply ignore.

There are three common ways for standards to be set: the marketplace, negotiation, and a standards leader (Varian et al. 2004). First are standard wars conducted in the marketplace. A classic example of this is the VHS and Betamax technology battle (Hall 2005) replayed today in Blu-Ray versus DVD. The benefit of allowing markets to set standards is that it follows the path of least resistance in terms of existing innovation. Technology flows where the skills and the markets already exist. Governments are not required participants in setting standards in this manner, but clearly, the legal, IP, innovation, and trade environments all play a part in determining how markets operate, and these are all areas in which governments set the tone. For countries that have national champions, allowing the markets to set standards can be tricky, as there is no guarantee that the home country's firm will survive the standard war. Nor is there a guarantee that consumers will direct the market to the best possible outcome; small events at different stages of development of a technology and a market can decide the outcome of a standards race, regardless of which product is technically superior (Arthur 1989; Malerba et al. 1999; Gallego 2010).

The second channel for setting standards is negotiation. Governments can play a direct role here. Recall the example of wireless signal security and China, and how it was resolved with international negotiations (primarily the US government). Negotiation typically involves compromise, which means that all players may have to sacrifice some element of gain for the betterment of the whole. Consumers can be the big losers in these situations as negotiations might not settle on the most cost effective or well-functioning standard. It is also difficult to determine exactly when a standard would be set naturally (Jakobs 2000) leading to the possibility that the necessity of entering into negotiations artificially sets the limit to further standard development.

Having a standard leader is the third form of standard setting. Such a leader can come out of a market fight, or by virtue of being the first to develop and disseminate a technology. Those wanting to supplant the standard may face extreme barriers to entry erected simply by the standard having been in place for a long period of time, or requiring a vast infrastructure that is infeasible to duplicate for a new standard (Gallego 2010). Perhaps no greater example exists of this than the Internet.

Standards Case—Brazil's Personal Computer Endeavor

In 1985, Brazil passed the National Information Technology Policy in an attempt to turn its burgeoning, domestic IT industry into a pillar of productivity and growth. The law blocked imports of some foreign computer and IT-related goods and for those imports that were permitted; foreign firms were required to interact with Brazilian-owned firms for in-country sales. This protectionist move had two goals: boost domestic technology growth by keeping out foreign competition; and provide an avenue for domestic economic development.

At the time of the passage of the Informatics Law, as it was called, Brazil did have a small computer manufacturing sector. Most of the domestically produced computers followed international standards and were clones of foreign market-leaders, while peripherals and software tended to have a higher degree of local content. The Informatics Law shifted all of the standard-setting to domestic producers.

Of the three methods for setting a standard—marketplace, negotiation, a standard leader—at the time of the passage of this law Brazil could not meet any. The market was not large enough to be self-sustaining, the protectionist move was unilateral and involved no negotiation with other countries or multilateral bodies, and Brazil possessed no domestic producer that was already a standard leader.

Brazil did see a growth in the domestic technology capability of some producers as the vacuum of foreign goods was filled. However, consumers suffered as the Brazilian products generally were more expensive and less reliable than their foreign competition. By the end of the 1980s, policymakers saw how countries like Taiwan and Korea were enjoying booming IT growth through much more liberal trade policies. Consequently, the Informatics Law was changed to allow for more foreign competition in the IT sector, and to de-emphasize the need for Brazil to set internal standards in that sector. Today, decentralized knowledge spillovers, as opposed to protectionist standards, are credited with boosting the IT growth of Brazil.

Sources: Botelho and Smith 1985; Perini 2006; Magalhaes et al. 2009

For developing country practitioners, care must be taken when agreeing to standards. Standards can be a benefit to a country's innovation efforts by providing guidelines for entrepreneurs entering the international market. Standards also can inhibit local innovation by preventing entrepreneurs from selling their products on the global market (Gibson 2007; World Bank 2010). The OECD recently compiled a review of empirical assessments of the impact of standards on international trade. Importantly, the studies examined found a mix of positive and negative impacts of both national and international standards on the conduct of international trade (Swann 2010). For a fitting example of variable impacts of a standard, one needs look no further than the well-documented controversies associated with one of the most important modern efforts at international standardization, that of IP under the TRIPS agreement.

7.7   Conclusion

Intellectual property protection is critical for innovative, knowledge-based firms. A company must be able to exclude others from using its research and inventions or else it will not be able to capture all of the value from its intellectual property. When a company is unable to exclude others, most commonly accomplished by enforcing a patent, another company can profit from using an invention without bearing any of the costs of the invention's development. The creator of the invention then becomes an unwilling subsidizer of a competitor.

A country must have enforceable intellectual property rights to protect the money and resources a company invests in research and development, or else investors in the company will not want to run the risk of having a knowledge-intensive company's most valuable asset, its intellectual property, stolen by a competitor. This both devalues the assets of knowledge- intensive companies and lowers the potential for profits. Weak property rights increase risk and discourage investment.

Policymakers must consider the impact of standards on domestic innovation. In the same manner as IP legislation discussed earlier in this chapter, any decision on standards must be made in light of factors such as the level of domestic innovation, domestic technology appropriability and usability (absorptive capacity), specific areas of technological strengths and weaknesses, and areas of potential trade growth.

References

Abel, A., Mankiw, N. G., Summers, L., & Zeckhauser, R. (1989). Assessing dynamic efficiency: Theory and evidence. *Review of Economic Studies, 56,* 1–20.

Arrow, K. (1962). Economic welfare and the allocation of resources for invention. In Richard R. Nelson (Ed.), *The rate and direction of inventive activity.* Princeton: Princeton University Press.

Arthur, W. B. (1989). Competing technologies, increasing returns, and lock-in by historical events. *The Economic Journal, 99*(394), 116–131.

Arvanitis, R., & M'henni, H. (2010). Monitoring research and innovation policies in the Mediterranean Region. *Science, Technology and Society, 15,* 233 doi:10.1177/097172181001500204.

Bird, R., & Cahoy, D. (2007). The emerging BRIC economies: Lessons from intellectual property negotiation and enforcement. *Northwestern Journal of Technology and Intellectual Property, 5,* 3.

Bizri, O., Chung, S. C., Kim, J. S., Larsen, K., Rischard, J. F., & White, J. (2010). *A short note on Lebanon's national innovation policy and system.* Draft White Paper. World Bank.

Botelho, A., & Smith, P. (1985). The computer question in Brazil: High technology in a developing society. MIT Center for International Studies: Paper Series.

Bramley, C., & Kirsten, JF. (2007). Exploring the economic rationale for protecting geographical indicators in agriculture. *Agrekon, 46*(1), 69–93.

Branstetter, L., Raymond, F., & Fritz Foley, C. (2005). Do stronger intellectual property rights increase technology transfer? *Empirical Evidence from U.S. Firm-Level Panel Data. Quarterly Journal of Economics, 121,* 321–334.

David, P. A. (1992). Intellectual property institutions and the Panda's Thumb: Patents, copyrights, and trade secrets in economic theory and history. In global dimensions of intellectual property rights in science and technology by the National Research Council. Washington, DC: National Academy Press.

DiMasi, J., Hansen, R., & Grabowski, H. (2003). The price of innovation: New estimates of drug development costs. *Journal of Health Economics, 22*(2), 151–185.

ESTIME. (2011). No date provided. *Science and technology and innovation profile of Jordan.* http://www.estime.ird.fr/IMG/pdf/Final_report_Jordan_IM_RA.pdf. Accessed 29 Dec 2011.

European Community. (1996). *Standardization and the global information society: The European approach.* White Paper. http://eur-lex.europa.eu/LexUriServ/LexUriServ.do?uri=COM:1996:0359:FIN:EN:PDF. Accessed 29 Dec 2011.

Fink, C., & Maskus, K. (Eds.). (2005). *Intellectual property and development lessons from recent economic research.* Washington, DC: The World Bank.

Gallego, B. (2010). Intellectual property rights and competition policy. In Carrea, C. Northampton (Ed.), *Research handbook on the protection of intellectual property rights under WTO rules.* Massachusetts: Edward Elgar.

Ganslandt, M., Maskus, K., & Wong, E. (2005). Developing and distributing essential medicines to poor countries: The DEFEND proposal. In Fink and Maskus (Ed.), *Intellectual property and development lessons from recent economic research.* Washington, DC: The World Bank.

Gibson, D., & Smilor, R. (1991). Key variables in technology transfer: A field-study based empirical analysis. *Journal of Engineering and Technology Management, 8,* 287–312.

Gibson, C. (2007). Globalization and the technology standards game: Balancing concerns of protectionism and intellectual property in international standards. *Berkeley Technology Law Journal, 22,* 1403–1484.

Godin, B. (2005). The knowledge-based economy: Conceptual framework or buzzword? *The Journal of Technology Transfer, 31*(1), 17–30.

Gort, M., & Klepper, S. (1982). Time paths in the diffusion of product innovations. *Economic Journal, 92:* 630–653.

Hall, B. (2005). *Government policy for innovation in Latin America.* Report to the World Bank, presented at the Barcelona Conference on R&D and Innovation in the Development Process, June 2005.

Harris, R. (2001). The knowledge-based economy: Intellectual origins and new economic perspectives. *International Journal of Management Reviews, 3*(1), 21–40.

Higher Council for Science and Technology for the Kingdom of Jordan. (2004). *Strategy for the higher council for science and technology, 2005–2010.* http://hcst.gov.jo/wp-content/uploads/2011/04/Strategy-2005_20101.pdf. Accessed 29 Dec 2011.

International Intellectual Property Alliance. (2009). *Jordan: IIPA special 301 report on copyright protection and enforcement.* White Paper. http://www.iipa.com/rbc/2009/2009SPEC301JORDAN.pdf. Accessed 29 Dec 2011.

International Intellectual Property Alliance. (2011). *Lebanon: IIPA special 301 report on copyright protection and enforcement.* White Paper. http://www.iipa.com/rbc/2011/2011SPEC301LEBANON.pdf. Accessed 29 Dec 2011.

IPR Commission. (2002). *Integrating intellectual property rights and development policy.* Commission on Intellectual Property Rights. http://www.iprcommission.org/papers/pdfs/final_report/CIPRfullfinal.pdf. Accessed 7 May 2012.

Jakobs, K. (2000). *Information technology standards and standardization: A global perspective.* Hershey: Idea Group, Inc.

Javorcik, B. (2005). The composition of foreign direct investment and protection of intellectual property rights: Evidence from transition economies. In Fink and Maskus (Ed.), *Intellectual property and development lessons from recent economic research.* Washington, DC: The World Bank.

Lall, S. (2003). Indicators of the relative importance of IPRs in developing countries. *Research Policy, 32,* 1657–1680.

Magalhaes, D., Knight, P., & De Costa, E. (2009). Will the Soccer World Cup of 2014 Help Bridge the Social Gap through the Promotion of ICT and E-government in Brazil? In S. Dutta & I. Mia (Eds.), *The global information technology report: 2008–2009.* Geneva: World Economic Forum.

Malerba, F., Nelson, R., Orsenigo, L., & Sidney, W. (1999). 'History-Friendly' models of industry evolution: The computer industry. *Industrial and Corporate Change, 8*(1), 3–40.

Malpani, R. (2009). All costs, no benefits: How the US-Jordan free trade agreement affects access to medicines. *Journal of Generic Medicines, 6*(3), 206–217.

Mansfield, E. (1985). How rapidly does new industrial technology leak out? *The Journal of Industrial Economics,* XXXIV(2), 217–23.

Mansfield, E., Schwartz, M., & Wagner, S. (1981). Imitation costs and patents: An empirical study. *Economic Journal, 91,* 907–918.

Ministry of Economy and Trade for the Republic of Lebanon. (2011). Services. http://www.economy.gov.lb/index.php/subSubcatInfo/2/86. Accessed 29 Dec 2011.

Mowery, D., & Sampat, B. (2005). Universities in national innovation systems. In J. Fagerberg , D. Mowery, & R. Nelson (eds.), The Oxford handbook of innovation (pp.209–239). Oxford: Oxford University Press.

Organization for Economic Cooperation and Development. (2011). Intellectual assets and innovation: The SME dimension. OECD Studies on SMEs and Entrepreneurship, OECD Publishing. http://dx.doi.org/10.1787/9789264118263-en.

Park, W. G., & Lippoldt, D. C. (2008). Technology transfer and the economic implications of the strengthening of intellectual property rights in developing countries, oecd trade policy working papers, No. 62, OECD Publishing. Paris, France.

Perini, F. (2006). The structure and dynamics of the knowledge networks: Incentives to innovation and R & D spillovers in the Brazilian ICT Sector. SPRU 40th Anniversary Conference, 2006.

Powell, W. W. & Snellman, K. (2004). The knowledge economy. *Annual Review of Sociology, 30,* 199–22.

Rischard, J. F., White, J., Chung, S., & Kim, J. S. (2010). *A candid review of jordan's innovation policy.* White Paper. World Bank and Korea RIAL.

Romer, P. (1996). Why, indeed, in America? theory, history, and the origins of modern economic growth. *American Economic Review 86*(2), 202–06.

Ryan, M. (1998). *Knowledge diplomacy: Global competition and the politics of intellectual property.* Washington, DC: The Brookings Institution.

Scotchmer, S. (2005). *Innovation and Incentives.* Cambridge, MA: The MIT Press.

Swann, G. P. (2010). International standards and trade: A review of the empirical literature, oecd trade policy working papers, no. 9. OECD Publishing.doi: 10.1787/5kmdbg9xktwg-en.

Tseng, C. (2009). Technological innovation in the BRIC economies. Industrial Research Institute Inc. 0895-6308/09.

UNCTAD. (2003). Technology transfer and intellectual property rights: The korean experience. Unctad-ictsd project on iprs and sustainable development. Issue 2.

Varian, Hal, Joseph Farrell, & Carl Shapiro. (2004). *The economics of information technology*, Cambridge: Cambridge University Press.

World Bank. (2010). *Innovation policy: A guide for developing countries.* Washington, DC: The World Bank.

World Bank. (2012 - unpublished). *Innovation policy handbook.* In N. Vonortas & A. Aridi (eds.), Washington, DC: World Bank Publications.

World Intellectual Property Organization. (2007). *WIPO patent report.* Geneva: WIPO.

World Intellectual Property Organization. (2011). *World intellectual property report: The changing face of innovation.* Geneva: WIPO.

World Intellectual Property Organization. (2011a). *Summary table of membership of the World Intellectual Property Organization (WIPO) and the Treaties Administered by WIPO, plus UPOV, WTO and UN.* http://www.wipo.int/treaties/en/summary.jsp. Accessed 29 Dec 2011.

World Intellectual Property Organization. (2013). *World intellectual property indicators.* Geneva, Switzerland: WIPO.

World Trade Organization. (2011). *Understanding the Wto: The organization members and observers.* http://www.wto.org/english/thewto_e/whatis_e/tif_e/org6_e.htm. Accessed 29 Dec 2011.

Wunch-Vincent, S. (2013a). The economics of copyright and the internet: Moving to an empirical assessment relevant to the digital age, economic research working paper #9. world intellectual property organization. Geneva: WIPO.

Made in the USA
Middletown, DE
25 September 2018